LIGHT MEAL RESTAURANT
DINE IN CULTURAL SPACE

(英)布兰登·海斯/编 鄢格/译

轻食餐厅

辽宁科学技术出版社

Contents 目录

006 **Preface: When Culture Meets Catering**
序言：当文化遇见餐饮

016 **Chapter One: Culture & Dining**
第一章：文化与餐饮

018 *1.1 Origin and Development*
起源与发展

022 *1.2 Pre-planning*
预先计划

026 **Chapter Two: Dining in Museums**
第二章：博物馆中的餐饮设计

028 *2.1 General Design*
总体设计

028 **2.1.1 Services Design**
服务设计

030 **2.1.2 Exhaust Design**
排烟设计

033 *2.2 Functional Design*
功能设计

033 **2.2.1 Café Design**
咖啡厅设计

034 Integrated Planning
整体规划

035 Menu Selections
菜单选择

037 Space Planning
空间规划

037 Equipment Design
设备设计

Display Design 陈列设计	038
Price Decisions 价格设定	039
2.2.2 Restaurant Design 餐厅设计	**040**
Circulation Diagram 动线规划	040
Location Selection 选址	040
Seating Capacity 座位容量	041
Considering the Queue 排队问题	042
Kitchen Design 厨房设计	043
Food Storage 食物储存	043
Waste Disposal 垃圾处理	043

Case Studies
案例赏析

West Valley Art Museum Café Renovation 西部峡谷美术馆咖啡厅翻新	046
Untitled 未名咖啡厅	054
Nerua Restaurant Nerua 餐厅	060
The Wright 赖特餐厅	068
Holburne Garden Café 赫尔本博物馆花园咖啡厅	076
Groninger Museum Restaurant "Mendini" 格罗宁根博物馆"门迪尼"餐厅	084
MOSI 曼彻斯特科学工业博物馆咖啡厅和餐厅	088
Coach House, Hatfield House 门房改造餐厅	096

104 **L'Osteria Künstlerhaus**
艺术之家活动中心 L'Osteria 意式餐厅

108 **The Whitechapel Art Gallery Dining Rooms**
白教堂画廊餐厅

114 *Chapter Three: Dining in Theatres*
第三章：剧院中的餐饮设计

116 *3.1. General Design*
总体设计

116 *3.2. Design Requirements*
设计要求

117 *3.3. Functional Design*
功能设计

117 **3.3.1 VIP Lounge / Boardroom**
贵宾室 / 会议室

118 **3.3.2 Bar (s)**
酒吧区

119 **3.3.3 Beverage Coolroom**
饮品冷藏室

119 **3.3.4 Beverage Store**
饮品存储区

120 **3.3.5 Service Kitchen**
服务厨房

120 **3.3.6 Coolroom**
冷藏室

121 **3.3.7 Freezer**
冷冻室

121 **3.3.8 Dry Store**
干货储藏区

121 **3.3.9 Catering Store**
餐饮用具存储区

122 **3.3.10 Staffroom**
员工区

3.3.11 Uniform Store 制服存放区	122
3.3.12 Staff Toilets and Changerooms 员工卫生间和更衣室	122

Case Studies
案例赏析

Public Theatre Library Lounge 公共剧院休闲酒吧	126
BELG AUBE Tokyo Metropolitan Theatre 东京艺术剧场 BELG–AUBE 小酒馆	134
The Swan at the Globe Theatre 环球剧院天鹅餐厅	140
Barbican Lounge 巴比肯艺术中心休闲餐厅	148
The Bar & Restaurant, Deventer Schouwburg 迪温特剧院酒吧与餐厅	156
The Grand Café & Brasserie Pollux, de Maaspoort Theatre Maaspoort 剧院咖啡餐馆	164
The Brasserie & Café, Theatre de Leest Leest 剧院咖啡馆	172
Phantom Restaurant 幻影餐厅	178
Canteen Covent Garden 科芬园餐厅	182
Café Bar Theatro 剧院休闲酒吧	190
Bar Agora, Theatre Modernissimo 集市酒吧，Modernissimo 剧院	196
Cinepolis Luxury Cinemas - La Costa Cinepolis 影院餐厅	202
Cinepolis Luxury Cinemas - Del Mar Cinepolis 剧院餐厅	208
Paard van Troje 特洛伊木马大厅	214

Index
索引 — 222

Preface:
When Culture Meets Catering

序言：当文化遇见餐饮

Preface: When Culture Meets Catering

Creating a dedicated overspill area for school parties at the Museum of Science & Industry, Manchester
曼彻斯特科学工业博物馆咖啡厅和餐厅内专为学生团体服务的就餐区

When Culture Meets Catering

In certain fundamental aspects, restaurants, Cafés and bars in cultural institutions are exactly the same as standalone hospitality spaces. Great food and great service will always be the fundamentals of any great recipe for success! In other ways, however, especially when it comes to branding and interior design concepts, these spaces can be complex and challenging to get right.

The main reason for this complexity is because of the difficult balance designers need to create between two spaces that are inter-related, but which serve very different functions. How does that relationship work in terms of branding and the look and feel of the space, for example, and what are the pitfalls that potentially lie ahead?

At the very beginning of the project, the designers of the hospitality space need to get to know the overall strategy of the cultural organisation. The synergy between the main content of the institution and its aspirations for its hospitality offer need to be explored via thorough questioning before moving forward. What is the space trying to achieve and what criteria is it fulfilling? Over-arching management plans to attract different demographic targets, for example, or to extend opening hours, will impact on the hospitality offer too, whilst a new brand vision to create a more cutting-edge space or one with more populist appeal would also create a different context for any dining area.

Target customer groups can be very varied – much more so than for a standalone restaurant, creating complex demands on a space that may need to house romantic dinners for couples in the evening and offer snacks for huge parties of schoolchildren during the day. Sometimes there is the possibility of being able to create flexible overspill areas for school parties or mothers with young children, where particularly robust furniture is a must, but which can be shut off for evening events or else moved round thanks to flexible furnishing arrangements.

当文化遇见餐饮

从基本功能上来说，文化机构中的餐厅、咖啡厅和酒吧与独立的餐饮服务空间完全一致。美味的食物和贴心的服务永远是成功的关键。但涉及到其他方面，如品牌定位和室内设计，这些空间则稍显复杂且更具挑战性。

其复杂性主要存在于设计师如何在两个功能不同但又相互联系的空间中取得平衡，这是一个难题。两个空间之间的联系如何通过品牌定位、外观风格和环境氛围体现出来以及确定存在哪些问题。

餐饮空间设计师需在项目初期了解文化机构的整体规划及理念。机构本身的主要功能和其对于餐饮服务的要求需全面而彻底地进行考虑——空间设计应满足何种要求以及须遵守哪些相关法则。吸引不同层次顾客的整体规划方案或者延长营业时间都会影响到餐饮空间的经营，而打造一个时尚流行的空间则会营造出与众不同的就餐环境。

与独立的餐厅相比，文化机构中的餐饮空间的目标顾客更加多样化，这就要求空间氛围的多样性——既适合情侣共进浪漫晚餐，又适合学生群体享用快餐。一些情况下，空间内可以预留出为学校团体或带孩子的妈妈们使用的区域。在这些区域中可移动家具是必需品，在晚上可以遮蔽或移走为举办聚会等活动提供空间。

序言　当文化遇见餐饮

The variety in the scope of customers doesn't mean, however, that designs should be bland or try to please everyone at the cost of having a dull identity. Cultural institutions who design their own hospitality spaces sometimes fail to create enough difference between their dining areas and their general public spaces.

Roy Westwood, Head of Innovation for Levy Restaurants, with whom we have worked on many restaurants within museums or arts centres, makes it very clear that "a restaurant space truly needs to be a restaurant and not a copy of a gallery or exhibition space. Clients often think that's the right way to go – to be a 'gallery that serves food' – but in fact you mustn't overplay the feel of the institution inside the dining space for the clear reason that people need a break from the intensity of what they are seeing in order to reflect on it all the better."

A restaurant always needs to function as a restaurant primarily and follow good restaurant design rules, but needs to have a definite relationship with the building it is housed in and the purpose of the institution – otherwise it goes too far in the other direction and looks unrelated and confusing for customers.

顾客类型的多样性并不意味着空间设计应平淡无奇——通过统一的风格迎合每一位顾客。在文化机构中，其餐饮空间往往与公共区域在风格上毫无差别。

我们曾同罗伊·韦斯特伍德，Levy餐厅创新设计引领者，共同设计多个位于博物馆或艺术中心内的餐饮空间。他一直强调："餐厅空间应具备餐厅风格，而不能模仿画廊或者展区。业主们通常认为文化机构中的餐饮空间应打造成'食品展示'区的风格。但实际情况是，游客需要的是一个游览参观之后可以小憩的空间，这样才更有助于让他们回味之前的体验。"

餐厅首要要行使餐饮空间的主要功能，遵循合适的餐厅设计法则。与此同时，还要与其所在的建筑和其功能构筑一定的联系。否则，太过脱离建筑本身则会让顾客感到迷惑。

Dining spaces need to provide contrast from gallery spaces, such as here at the Barbican Lounge, so that visitors can reflect on what they have seen
巴比肯艺术中心休闲餐厅内就餐空间和画廊风格形成鲜明对比，便于游客识别

Preface: When Culture Meets Catering

When it comes to branding, the space can't be unbranded and anonymous, or else it will risk looking like an afterthought, but the branding and positioning have to work with and not against the existing overall brand and its position. For example, when we created the new restaurant spaces for leading British arts institution the Barbican Centre, we had to respect the Centre's strong identity and its rulebook on graphics and fonts, but still have a new playful identity.

Once that relationship has been explored, the design team can move on to exploring the concept, asking what it is trying to achieve and what statement it is trying to make. Will it have a celebrity chef for example and what is its basic identity (café, bar, fine dining, casual dining)?

Whilst context is everything, the relationship between the two spaces needs to remain subtle and un-forced. When we designed both a restaurant and a café space for MOSI (the Museum of Science and Industry) in Manchester, we created huge-scale steel servery tables that were inspired by laboratory benches, subtly referencing the museum's industrial context.

提及品牌推广，文化机构中的餐厅在定位上与其所处的机构相互融合与一致的同时，应具备自身的特色。举个例子，我们在芭比肯艺术中心内设计全新的餐饮空间时，在尊重中心自身的浓郁风格并遵循其在图形和字体方面的要求的同时，赋予餐厅更多的趣味性。

确定某种关联之后，设计团队便可以开始探索设计理念，考虑应达到某种效果以及如何实现。例如，是否聘请知名大厨以及定位在某种风格（咖啡厅、酒吧、高档餐厅或快餐厅）。

背景环境决定一切，餐饮空间和其所处的文化机构之间应保持一种微妙而互惠的关系。例如，我们在曼彻斯特科学与工业博物馆小餐厅的项目中，专门打造了规格较大的钢结构餐桌。其灵感主要源于实验室中使用的桌椅，微妙地与博物馆的工业背景环境联系起来。

Servery tables inspired by laboratory benches at the Museum of Science & Industry in Manchester
曼彻斯特科学工业博物馆咖啡厅和餐厅内模仿实验室长凳而设计的就餐桌

Branding for the new Barbican restaurants is referential but playful
巴比肯艺术中心休闲餐厅形象设计趣味十足

序言　当文化遇见餐饮

Wood from the local estate is used throughout at the Coach House restaurant, Hatfield House
哈特菲尔德庄园餐厅内采用的木材从当地旧建筑中回收而来

For the award-winning Coach House restaurant, set within historical Hatfield House (where Queen Elizabeth I spent much of her childhood), we used English suppliers and English materials to link thematically to the main attraction, even using wood from trees felled by the estate's foresters, with the stumps used as support plinths for external hanging signs, and dressed wooden blocks for internal signage and menus. For the Barbican Lounge, we created a bespoke resin floor in a wonderful peacock green/blue, colour-matched to a photo of the outside lake taken in the summer months.

The dining space also needs to be designed to compete with what is happening outside on the high street. Dining in a cultural institution is often a very spontaneous decision and visitors may decide on the spur of the moment if they are eating in or eating out. Just because the museum or theatre restaurant is closest, this does not mean it will necessarily be a first choice, so visitor custom can never be taken for granted. If you can entice customers to stay in to eat, though, profit is of course increased for the institution and also visitors are more likely, once refreshed, to go back into the wider space for more viewing and possibly some purchasing in the shop – especially if encouraged by a glass of lunchtime champagne!

"Clients also tend to think of dining customers as single users," Roy Westwood also commented, "but a catering offer can play a real role in converting customers who will then use the restaurant as a destination for social or even business meetings and can even play a role in people's choice between visiting one major museum or another, for example." A return customer can then be informed about future events and this knock-on chain of events may mean a new visit to the institution might result. The power of a great space within a cultural institution, for owners, operators and designers that get it right, cannot be underestimated!

Brendan Heath
SHH Associate Director

我们曾经负责设计位于哈特菲尔德宫（伊莉莎白一世曾在这里度过大部分的童年时光）内的 Coach House 餐厅。我们与英国供应商密切联系并大量运用英国本地的材质，以与餐厅所处的背景环境融合。例如，使用的木材是庄园内部种植的，木墩用做支撑室外标识的底座，经装饰的木块用于制作室内标牌和菜单板。在芭比肯酒吧中，我们专门打造了孔雀绿色的树脂地面，与一幅在夏季拍摄的室外湖泊照片相互呼应。

餐饮空间的设计同时应考虑到外部环境。在文化机构中就餐通常是很随机的决定，游客往往在一瞬间而决定。尽管餐厅距离博物馆或剧院较近，但这并不意味着这里的餐厅就是就餐的第一选择，因此不要理所当然地将游客视为固定的顾客。如果能够吸引游客进入就餐，则会相应地提高文化机构的收入——游客在稍事休息之后，可能会再次回去参观或者进到商店购买纪念品。当然，如果能够提供一杯香槟，那么顾客更会被深深地吸引。

罗伊·韦斯特伍德曾提出说："业主通常将就餐顾客视作一次性消费人员。其实不然，供应的食品和提供的服务扮演着重要角色，完全可以吸引顾客来到附近参加社会活动或开会时再次前来就餐，即便那些再次来博物馆附近参观的游客也可能成为潜在的回头客。"顾客可以被告知未来时间内文化机构将举办的活动，如此一来，可以增加其参观次数，从而获得收益。文化机构内餐饮空间若设计得当，其带来的利益对于业主、经营者和设计师来说是不可预估的！

SHH 室内设计公司副董事
布兰登·海斯

Preface: When Culture Meets Catering

Catering facility in the Deventer Schouwburg: the adage: 'see and be seen'
迪温特剧院酒吧与餐厅独特的就餐设备——"见与被见"

The theatre world stands on the verge of great challenges. How can the theatre safeguard its position in the world of leisure and entertainment and remain appealing to its audience? The selection of leisure time activities is going through impressive changes; globalisation and digital developments are impacting our society enormously. Internet is the doorway to a world that is becoming increasingly 'smaller'. We have access to the worldwide web whenever and wherever we want and have news, information and entertainment at our fingertips. This means that entertainment is always available at our own convenience. People are no longer bound by theatre show starting times, for example. This provides enormous freedom, since every individual can decide to take relaxation when, in whatever form and at the hour they want. This individualisation also brings new challenges.

Given that every individual can arrange their lives around their choice of leisure time activities for example, it creates the risk that people will become isolated. Every change creates new developments; action is reaction. Of course, this freedom of choice in leisure activities presents countless possibilities. After all, humans are social creatures. It is precisely the interaction with other people that is important for the development and the well-being of the individual. For humans, partnerships or marriage and having a family are crucially important to happiness. Joint activities and especially friendships are deemed, by the majority of people, to be the most important factors for happiness. So as well as freedom of individual choice, participating in group experiences is enormously important.

That is why the cultural world, in which the theatre performs an important role, should certainly respond to those needs. Theatres should offer more than just the facilitation of the programming.

Museums with window displays in the façade, theatres with transparent rehearsal rooms and cultural buildings where the partitions between the interior and exterior are eliminated as far as possible and where people would happily spend the whole day.

影剧院的经营目前面临着严峻的挑战，如何确保其在休闲娱乐领域的地位并吸引更多的观众亟须考虑。休闲时间活动的选择发生了巨大的变化；国际化以及数字化在很大程度上影响着社会的发展。如今，互联网已经成为了通往世界的"门户"。现今，我们可以随时随地通过互联网浏览新闻、信息和进行各种娱乐活动。这就意味着我们再也不必到影剧院中休闲娱乐。每个人都可以根据自身需求选择时间、地点和形式进行放松娱乐。这在赋予我们更多自有的同时，也带来了新的挑战。

每个人都根据自己的需求选择娱乐活动，如此一来便阻断了人与人之间的交流与联系。每一次变化都会带来全新的发展，行动与反应往往相辅相成。不可否认，自由选择娱乐活动呈现出多种可能性。但人类毕竟是社会性动物，人与人之间的互动对于个人的身心健康来说至关重要。对于人类来说，婚姻和家庭是幸福的重要元素。集体活动和友情更为大众需要。因此，个人选择性的自由与集体体验同等重要。

影剧院在文化世界中占据重要的地位，应满足观众的需求，举办更多的活动。

博物馆外观上的陈列橱窗和影剧院内通透的排演室意味着文化建筑中室内外界限的逐步摒弃，人们可以在这里度过愉快的一天。

As it now stands, we have excessively tucked away art in impenetrable boxes, since 'theatre is performed in a black box and visual art hangs on snow white walls.'

Theatres should not be impenetrable establishments but should be more a continuation of public spaces; with low-thresholds and various facilities appealing to the needs of people seeking entertainment. 'You must of course acknowledge that theatres and pop music need black boxes, places for the concentration that performances or concerts require.' As architects, we hold the opinion that the surrounding spaces and the connection to the rest of the city should be designed in such a way that they become exciting places that evoke curiosity about what's going on inside.

我们习惯于将艺术品陈列在坚不可摧的盒子中展出，然而戏剧节目在"黑匣子"中表演，艺术品陈列在雪白的墙壁上已经成为了一种新兴的模式。

影剧院不应该是一个封闭的场所，应该是公共空间的延续。只有放低门槛，增加多样的娱乐设施才能够吸引更多的观众。作为设计师，必须明确戏剧和流行音乐需要在"黑匣子"（表演或音乐会需要的空间）中表演。我们坚信周围的空间以及与整个城市的联系应该以这样一种方式打造——一个令人兴奋的空间，让人期待里面正在上演的一切。

Transparent façade of the Chassé Theatre Breda, The Netherlands
芬兰 Chassé 剧院通透的外观

Preface: When Culture Meets Catering

Experience restrooms as a part of the theatre concept
卫生间作为剧院设计的一部分

Brasserie light fixtures positioned above the bar turning slowly
吧台上方的灯饰不断地旋转

Catering plays a significant role within this philosophy and can contribute to the distinctive positioning of the theatres; by using hospitality formulas that are aligned to the theatres and that function as additional, unique selling propositions. In turn, the hospitality formulas can make use of the positioning of the theatres; a complete, mutually reinforcing concept.

We recognise these opportunities in our theatre architecture and strive to bring the experience and emotions from within the black box to the forefront, visible and accessible. Seeing and being seen is the motto for theatres and their hospitality formulas. Going out for dinner is one of the many possible leisure activities and what could be more appealing than being part of the cultural programme. A cabaret performer in The Netherlands once said: "More theatre is performed in the foyers than on the boards." By utilising the ambiance and qualities of the theatres and giving the food and beverage formulas well-considered interpretations, the catering and the theatres will reinforce each other.

In the design for the Brasserie for the Chassé Theatre in Breda we used the theatre's added values and applied them to the interior architecture. The Brasserie has a centrally placed bar, positioned as a centrepiece in the space and with seating grouped around it.

Part of the space is raised so it can also be used as a stage. The theatrical aspect has been translated into a variety of light fixtures positioned above the bar turning slowly until they disappear into the pantry, so the space is continually changing. The light fixtures are the props in the theatre show performed by the catering employees and guests.

Another challenge is using the theatre buildings more intensively. So not only when there are performances but also at other times. The theatres can benefit in this way by opening during the daytime and making their hospitality facilities available. The foyers could be locations for corporate presentations, debates and also, predominantly, as open stages where upcoming talents can present themselves.

餐饮在这一设计理念中起到至关重要的作用，同时对于影剧院的定位起到帮助。餐饮服务模式融入到影剧院的经营中，并使其成为独特的营销点。反过来，餐饮服务也可充分利用影剧院自身的定位，打造一个全面而互补的理念。

我们在影剧院建筑项目中找到了机会，努力将黑匣子内的体验和氛围呈现出来。"见与被见"是影剧院服务模式的主要理念。外出就餐是众多休闲娱乐活动方式之一，而置身在文化环境中就餐则更具吸引力。一位来自荷兰的表演艺术家指出："如今越来越多的影剧院摒弃了舞台，乐于将表演安排在大厅中。"充分利用影剧院特有的氛围和品质，优化餐饮模式，从而实现两者的完美结合。

在位于布雷达的Chassé剧院小餐馆项目中，我们充分利用了剧院的附加值，并将其运用到室内空间设计中。小餐馆中设置着一个中央吧台，构成空间的中心，座位围绕在四周。

部分空间被抬升，可用作表演舞台。吧台上方悬挂着形状各异的灯饰，不断旋转并一直消失在备餐室中，整个空间的氛围在不断地变换着。这些灯饰是舞台的重要元素，如今成为了员工和顾客"表演"的烘托。

面临的另一个挑战是如何充分利用影剧院建筑空间，包括表演时间之外建筑的使用。影剧院可以在日间开放，提供便利的服务并从中获益。大厅可以用作举办演讲和辩论赛，同时可以用作开放式的舞台供相关人员表演使用。

序言　当文化遇见餐饮

Front cooking: positioning the chef centrally at the bar
表演烹饪：大厨成为主角

Ticketbox at Chassé Theatre
Chassé 剧院售票处

Theatres can offer opportunities for students and the corporate world to get together, hold meetings or share knowledge in, for example, 'seats to meet' types of settings. Theatres should facilitate fast wireless internet and become special, inspiring meeting places. Theatres are generally easily accessible, centrally located and have sufficient parking facilities. In short, offering ample opportunities.

Catering can perform an important role, for example with breakfast and lunch provisions. The challenge lies in choosing a well-thought-out product mix, distinctive from other venues. For example, by using unexpected combinations, nothing too complicated but certainly special. For the food and beverage concept for the expansion of the foyer in the Chassé Theatre they opted for a food experience by positioning the chef centrally at the bar: front cooking.

The dishes are various snacks inspired by 'street food' that can be bought on streets around the world. All dishes are offered at the same price. The various dishes hang on the lamps in the interior.

The theatre world will have to change. Art has become alienated from society, or literally: we have hidden art away in black and white boxes. If you improve the architecture, the public will feel more connected and proud of the culture created within. I believe that a large proportion of people's engagement is determined by buildings and their appearances. The total concept: whereby the architecture, the programming and use of the theatres and the hospitality formulas are all components for determining success. Unique, distinctive and appealing to a broad audience. An apparent contradiction, however, lies precisely in the challenges within the theatre world for architects, designers and food & beverage concept creators, 'when culture meets catering'.

Hans Maréchal ｜ M+R interior architecture – The Netherlands

影剧院可以为学生和企业员工提供交流或聚会的空间，如举办会议等。影剧院内应配备无线网路装置，便于会议期间使用。此外，影剧院应选址在交通便利的区域内，并确保具备足够的停车空间。总之，就是要提供一切可能的机会。

餐饮空间发挥着重要的作用，提供早餐和午餐服务。其面临的主要挑战便是如何配备食品结构，使其与其他空间食品供应大为不同。例如，尝试各种不同的食物组合，不求复杂但力求特色十足。在Chassé 剧院小餐馆中，设计理念是打造与众不同的就餐体验——大厨在吧台中间现场制作食物。

小餐馆内供应品种多样的零食，其灵感源自于"街边小摊"。这里所有的食物统一价格，食物悬挂起来供顾客选择。

影剧院的发展不得不面临着改变，艺术已经在逐渐地与社会脱离。这也就是说，我们将艺术隐藏在了建筑背后。如果我们能够改变建筑样式，公众便会感觉到其与艺术的联系，并为置身于文化氛围中而感到骄傲。我一直坚信，公众的参与程度很多取决于建筑本身以及其呈现的形式。总体的理念即为，表演、影剧院的使用方式以及服务模式是决定其能否成功经营的所有要素。独特而引人注目则必定会吸引大量的观众前来。一个显著的矛盾即存在于影剧院本身为建筑师、设计师和餐饮管理者带来的挑战。这就是"当文化遇到餐饮"。

汉斯·马利夏 / 荷兰 M+R 室内设计公司

Chapter 1:
Culture & Dining

第一章：文化与餐饮

1.1 Origin and Development

起源与发展

1.2 Pre-planning

预先计划

Chapter 1: *Culture & Dining*

1.1 Origin and Development

Food can play a key role in fostering relationships, building new audiences and creating financial sustainability for cultural institutions. As communities increasingly self-sort by politics, race, culture and income, food is one of the deeply human ways we come together and explore commonalities. The experience of growing food reconnects us to nature and fosters thoughtful awareness about what we eat. Preparing food helps us to share traditions and culture. Cultural institutions are embracing the fact that food strongly influences where and how we spend our time. Research on participation in the arts shows that while people are becoming less likely to partake of 'high culture' (museums, galleries, concerts and theatres), they increasingly attend multi-faceted cultural events that include food in the mix. Young people say an important aspect of a welcoming public environment is the ability to eat and drink with friends.

1.1 起源与发展

在文化场所中，食物扮演着重要角色，可以帮助促进发展，吸引公众，创造经济效益。社区往往通过政治、种族、文化和收入而划分，而食物则是让人们走在一起并探讨共性的方式。食物的发展历程将人类与自然联系在一起，唤醒人类的餐饮意识。食物制作帮助人类分享传统和文化，文化机构一直坚信食物可以影响人们对于场所的选择以及度过时光的方式。调查表明，公众对于单纯的博物馆、画廊、音乐会和剧院的光顾次数越来越少，更倾向于选择多功能并带有餐饮场所的文化机构。在年轻人眼中，一个受欢迎的公共环境就是一个能够与朋友吃喝的场所。

Ground floor glass extension of Coach House Restaurant in Hatfield House
哈特菲尔德庄园餐厅内扩建的玻璃结构就餐区

The new restaurant 'Nerua' opens its door in Guggenheim Museum Bilbao
古根海姆博物馆内新餐厅

第一章　文化与餐饮

The Library Lounge in Public Theatre has become a popular gathering place for New York people
公共剧院休闲酒吧已成为备受纽约人欢迎的场所

People could enjoy delicious food and good drinks during the show time in Deventer Schouwburg
人们可以在迪温特剧院酒吧与餐厅享用美味的食物

Food service in cultural institutions is nothing new. By the early 20th century, the presence of museum tearooms was expected. By mid-century the fashion was to have lunch spots or Caféterias. Beginning in the 1970s, people looked at restaurants as more than a rest stop for weary visitors. Many directors regard their restaurant facilities solely as a service for visitors and staff, and some may even consider it a necessary evil. A well-merchandised, well-operated restaurant can be a profitable venture and a continuing source of funds for operations.

Profitable, maybe, but not necessarily gourmet. The visitors historically are going in with an expectation that the food is not going to be great. They are trying to turn that perception around.

A generation ago, most cultural institutions would have considered it somewhat undignified to aggressively pursue catering business. Today, even facilities with modest attendance have discovered that these activities can offer a significant annual income stream, regardless of the economic climate.

文化场所中的餐饮空间并不是一个新鲜事物，在20世纪早期，茶室便已出现在博物馆中。到20世纪中期，提供午餐的场所或小餐馆开始备受期待。20世纪70年代初期，餐厅已不仅仅是为疲惫游客提供的休息场所，但多数仅为游客和内部员工使用。一个运营良好的餐厅可以带来可观的收益，为文化机构的运营提供资金支持。

无可否认，餐厅可以创造效益，但其提供的不一定是美食。之前，许多游客对这里提供的食物并不抱有很大的期望，而且这一想法已经根深蒂固。

约30年前，大多数文化机构认为在其中开设餐厅是一件不太体面的事情。然而在今天，即使游客较少的机构也发现开设餐厅可以带来可观的收入（不受当地经济情况的影响）。

Chapter 1: Culture & Dining

In the past, many of them allowed a wide variety of caterers to use their facilities because they believed that this policy removed a deterrent to rentals for those clients with a marked caterer preference. In practice, however, clients in search of alternate venues clearly select the venue first and the caterer second. Most cultural institutions believe that working with an exclusive caterer or maintaining a very short list of approved caterers leads to the most effective stewardship of the property. They also believe that these relationships offer an incentive to caterers to share revenue.

The issue always arises as to whether an exclusive caterer can provide the creativity and the range of price points to satisfy a diverse market. In part, this is a contractual question, dependent upon the institution's assessment of the resources within its marketplace. There are many off-premises caterers that can operate at a broad variety of price points, provided that the contract with the institution allows them to do so and that they receive clear direction from the client.

过去，许多人认为允许一系列的餐饮商直接进驻可以免去了租金对其造成的威慑。但实际上，业主通常是首先选择地点，其次再选择餐饮商。大多数文化机构认为同一家餐饮商合作便于管理，同时可以促进利润分配。

然而，这一经营模式带来的问题是独家餐饮商是否具备足够的创意和多样的价位，以满足多样化的市场需求。在一定程度上，这一问题源自该机构对于市场资源的评估。许多餐饮商可以提供灵活的价位服务，前提是合同规定中允许并得到业主的明确指示。

文化机构内的餐饮经营可以为其带来更多的收益，但同时必须保证文化机构具备较高的受欢迎程度。

Bar area design of Untitled in Whitney Museum of American Art
惠特尼美术馆内酒吧区设计

The famous Wright restaurant in New York Guggenheim Museum
古根海姆博物馆内赖特餐厅

第一章　文化与餐饮

Internal catering offers cultural institutions major benefits. It must be clearly understood that the client institution is the most favoured nation. The visitor's restaurant is an equally vital part of the equation, although it does not drive the income stream. Any cultural institution with a length of stay of an hour or more will find that visitors demand some type of refreshment. With a two-hour length of stay, some food service becomes essential to prolong the visitor experience. In fact, tour operators will often consider restaurant, café, or box lunch programmes as important elements in the development of their packages.

An effective restaurant can also bring significant ancillary benefits. Food service can increase the length of stay, which will enhance per capita spending in the shop and will be a factor in promoting membership. Restaurant discounts or member reservation policies can be part of the added value package for donors. With the exception of the highest attendance facilities, visitor food service, whether it be a snack bar, a café, or a table service restaurant, is a loss leader economically, although a superior café or restaurant may prove to be an important adjunct to the museum's development activities.

游客餐厅虽不能推动收益流的增长，但却是其中重要的一部分。不难发现，游客在文化机构内的停留时间达到或超过一小时，就会需要片刻的休息。如果停留时间多达两小时，那么餐饮服务就显得至关重要。旅游经营者通常将餐厅、咖啡厅等作为服务的一部分。

有效的餐厅经营模式往往能够带来更多的附加利益。餐饮空间可以延长游客停留的时间，从而提升在商店停留的人均时间，便于发展更多的会员。餐厅折扣或会员预定活动是提升附加值的一部分。除去那些游客光临次数最高的空间，游客餐厅服务（小吃店、咖啡厅、有餐桌服务的连锁餐厅）虽不能带来更多的经济效益，但却是文化机构中重要的组成部分。

The Clay furniture concept is brought into the design of Mendini restaurant in Groninger Museum
格罗宁根博物馆"门迪尼"餐厅内引入了陶土家具

Lighting is a key point in the design of Barbican Lounge in Barbican Centre
巴比肯艺术中心休闲餐厅内灯光设计是主要元素

Chapter 1: Culture & Dining

Typically, visitor food service is a loss leader from the caterer's perspective as well, although he or she may receive business from other events that are held after hours, elsewhere on the grounds, or in conjunction with local events. There are institutions that provide food and beverage service for themed candlelight dinners; car, boat, or antique shows; and concession stands at community festivals.

1.2 Pre-planning

It is extremely important to start planning early in the design process. For designers who have worked closely with many prestigious cultural institutions, they have learned how critical such early involvement is to the success in restaurant and restaurant planning. That measure of success is restaurant that enhances the public's visit and restaurant that delivers revenue to the institution.

All aspects of revenue generation should express the character of an institution. Of course, an institution's purpose is not to simply provide revenue, but to extend the experience, help make a memorable impression, and perhaps offer the visitor something to take home. Sensitive restaurant planning contributes to this by offering a place for reflection, absorption, and relaxation. Changing environments and varying light levels enhance the visitor experience.

在餐饮经营者看来，游客餐厅本身就是一种亏本的经营形式，但是他们可以通过其他形式获得收益，如举办活动等。有些文化机构专门为烛光晚餐的客人或者为车、船及各种表演等提供食物。

1.2 预先计划

在设计初期开始进行计划往往起到至关重要的作用，那些与知名文化机构密切合作的设计师非常了解早期介入对于餐厅成功运营的重要性。其中，餐厅的成功经营通常通过顾客的数量和其提交给文化机构的利润来衡量。

创收的所有方面能够诠释出一个机构的性质。当然，一个机构的经营目标不仅仅是简单地创作收益，更多的是营造一种体验，留下深刻的印象或者让游客思考。餐厅就是一个供游客思索和放松的场所。变幻的氛围和多样的照明方式帮助深化游客的体验。

The spiralling white design comes from the iconic exterior design of New York Guggenheim Museum
餐厅内蜿蜒的白色结构源自古根海姆博物馆的标识性外观设计

第一章　文化与餐饮

Thoughtful selection of an appropriate food menu further defines an institution's character. Not only is it important to develop a menu that is of interest to the audience, it is also vital to reinforce an institution's mission through an elegant presentation. The available menu and type of service has an enormous impact on planning requirements as it guides the type of kitchen and support one needs. Be it a café, food court, cart, or fine dining.

The real estate adage of 'location, location, location' applies to revenue generation as well. To maximise revenue retail, facilities must be conspicuous to visitors. Co-locating them in 'free-zones' allows for extended hours of operation and offer visitors the chance to browse through a shop after dining, before a lecture, or following a movie. Facilities can be designed in a way that does not compromise the importance of the interpretive message and a visitor's first impression of an institution.

The metal menu stands in Coach House Restaurant
哈特菲尔德庄园餐厅内金属菜单

People could relax and have a drink in the lobby of Theatre de Leest
人们在Leest剧院咖啡厅休闲放松

同样，一份合适的菜单能够进一步展现出机构的特质。菜单的选择不仅仅是要吸引顾客的注意，更需要通过优雅的展现方式强调机构自身的使命。无论是何种类型的餐饮空间（咖啡厅、美食广场），菜单和服务方式都会对规划要求带来深刻的影响。

Chapter 1: *Culture & Dining*

Dishes shelves
餐具柜

For larger institutions, multiple clusters of food and retail offerings not only offer choices to the visitor, but also may help capitalise on temporary exhibits or unique places within a building.

Special events, such as banquets and lectures, offer great potential for income to an institution and expose attendees to a facility they might not ordinarily visit.

Some questions must be considered: What is appropriate to your location and visitation? Are the facilities sized correctly to provide optimum revenue? What does my audience expect and what is the take-away message?

It is important to determine primary staff and vendor functions, such as the handling of incoming food products, garbage collection, and cash management. Limiting the access of food service staff, caterers, and vendors to public venues is preferred in order to prevent theft and damage to exhibits. Some other questions that should be addressed include: Does the staff require changing areas, lockers, and separate restrooms? How many staff will be required to support the service?

In the early phases of design, it is important to determine what type and size of utilities and engineering support will be required, such as water, power, gas and kitchen exhaust and make-up air. Many of these services should be crafted based on the anticipated menu and style of food that is to be served.

Other considerations are determining the hours of operation. Will the restaurant be available before and after the opening? Who will operate your facility – in-house or contracted? If contracted, what impact will this vendor have on the institution?

房地产的黄金法则也可促进创收。为提高零售收入，餐饮空间需选址在显而易见的位置。而将其同商店设置在一起可以延长营业时间，方便游客在就餐之后或者演讲之前闲逛。当然，这些空间的设计不能影响信息的传达以及机构给顾客带来的第一印象。

对于那些大型的机构而言，餐饮和零售空间不仅仅为游客带来更多的选择，同时还可以帮助举办临时展览等。

晚宴、演讲等特殊活动可以为机构带来更多的收入，同时为与会者提供特殊的体验。

规划过程中应考虑如下问题：如何确定合适的选址，空间规模是不是符合最佳效益原则，游客最期待的是什么。

确定员工和供应商的职责同样至关重要，如食物处理、垃圾收集以及现金管理等。应限制员工、餐饮服务商以及供应商进入公共场所的次数，避免展品的丢失和损坏。其他仍需考虑的问题包括：是否为员工提供更衣室、储物柜和单独的卫生间以及应雇用的员工数量。

设计早期，应确定机械设备的类型和规模（水、电、气、厨房排烟和补给空气）。所有这些应以菜单和食物风格而决定。此外，还应确定营业时间（文化机构每天营业之前和关闭之后餐厅是否继续经营）、餐厅选址（文化机构内部还是独立经营）。如采取独立经营的方式，应预估其对于文化机构的影响。

第一章　　文化与餐饮

Other questions to consider include: Will special events always be catered or will you provide the food? Who can you trust to take good care of and adequately clean up and leave no traces or damage to finishes? It is obviously important to provide restrooms, a catering pantry, public and loading access adjacent to the event area. Considerations also need to be given to permit pre-event set without disturbing daytime operations.

Advance planning of restaurant can yield long-term benefits for an institution. Starting planning in the earliest stage of schematic design helps a facility maximise revenue and minimise costs adding to the visitor experience.

其他仍需考虑的问题如下：特殊活动是否经常举办，是否需要供应食物、是否有合适的人员负责保管和清洁工作避免造成损失。卫生间、餐饮厨房、与活动区域临近的公共入口等都是至关重要的因素。当然，应确保活动准备工作不能影响文化机构的正常经营。

餐饮空间的提前规划能够为文化机构带来长久的利益，早期规划可以提升收入同时减少预算。

Restaurant restrooms design
餐厅内卫生间设计

Chapter 2:
Dining in Museums

第二章：博物馆中的餐饮设计

2.1 General Design
总体设计

2.2 Functional Design
功能设计

Case Studies
案例赏析

Chapter 2: Dining in Museums

2.1 General Design

2.1.1 Services Design

The services design applies to all facilities in museums: caterers, restaurants, Caféterias, take-out facilities at meeting venues, picnic baskets and catered lunchboxes. It should also apply to any restaurant services associated with package tour or related to museum activities.

The operators should make a commitment to the environment by avoiding disposable dishes and single-serve containers. However, as indicated, it is also important to have a commitment to reuse, recycle, and compost and to reduce the use of energy, water, and hazardous products in daily operations.

Dishes, cutlery and linens are reusable: avoid paper or polystyrene cups, paper napkins or tablecloths, plastic cutlery or disposable doilies, etc. When people are eating outside, use sturdy, reusable dishes (reusable acrylic) or compostable dishes. Ban disposable items such as plastic straws and stir sticks.

Cloth table napkins are more expensive at the outset, but they can be reused. When using cloth table napkins, be sure to look for organically produced cotton, hemp or linen items. This greatly increases the green aspect of the table napkin because no hazardous pesticides or chemical fertilisers were used in their production.

Serve tap water in reusable pitchers.

Avoid using single-serve containers for food and condiments (for example, milk, cream, sugar, artificial sweeteners, butter, ketchup, vinegar, mustard, jams, salt, pepper, and breakfast cereal). Provide these in bulk unless local health authorities prohibit it. (If for any reason health regulations prohibit it, convey the information to clients. The food service establishment should work with local health authorities to overcome

2.1 总体设计

2.1.1 服务设计

服务设计适用于博物馆内部所有类型餐饮方式，如宴席包办、餐厅、食堂、外卖、野餐篮、午餐便当等。同时，也适用于同旅行和博物馆活动相关的餐饮服务方式。

经营者应保证不破坏环境，如避免使用一次性餐具和包装盒。同时，应注重循环和回收利用，减少水、电以及危险物品的使用。餐盘、刀具等须循环使用，避免纸（聚苯乙烯）杯、纸巾、塑料刀具和一次性台布的使用。客人在室外就餐时，可为其提供坚固的餐具或环保餐具。禁止使用塑料吸管和一次性搅拌杆等。

布材质的餐巾虽然价格昂贵，但却可重复使用。当采用此种餐巾时，确保其是有机生产的棉、麻或亚麻材料制成。此类餐巾在加工过程中不会添加杀虫剂或化学物质，提升环保特色。

使用可重复使用的水罐供应自来水

避免使用一次性容器盛放食物和原料（牛奶、奶油、糖、人造甜味剂、黄油、番茄酱、醋、芥末、果酱、盐、辣椒、燕麦等）。除非当地卫生部门禁止，这些物质可成批供应（如果卫生条例禁止，则可向客户说明。食品供应机构可同当地卫生部门商讨，取消这些限制）。

regulatory hurdles.)

Whenever possible, provide beverages in bulk or in reusable or recyclable containers as listed in the facility's recycling programme (no juice boxes).

Make sure the facility provides recycling bins for glass, metal, and plastic items. Those recycling bins should be in visible locations, such as in kitchens and dining areas, and instructions provided should be clear.

Implement a system for collecting food waste: compost or supply livestock producers; Print menus on paper with a high post-consumer recycled content, preferably unbleached.

Try to buy food and beverages produced and sold locally (fruit, vegetables, meat, fish, poultry, wine, beer, etc.).

Buy fish and seafood that is Blue Ocean certified; Do not serve food grown, processed or preserved with ozonedepleting substances.

Buy organic food, wine and beer from sustainable farm sources; Offer vegetarian menus.

Give any usable surplus food to charities, if health regulations permit. Refuse to use products with excessive packaging.

Opt for reusable containers for transportation and delivery of food.

Make sure hazardous cleaning products are properly stored, used and disposed. Staff must receive proper training on the subject. Identify substitutes for cleaning products containing hazardous chemicals.

Consider the farmers' markets in local area.

如果可以，饮料也可盛放在可反复使用的容器内成批供应。

确保具备回收箱，用于盛放玻璃、金属及塑料物件。回收箱应放置在显眼的位置，如厨房和就餐区，标示应当足够清晰。

配备废弃食物回收系统；餐单上印上"消费后回收内容"。

购买当地生产的食物和饮料（水果、蔬菜、肉、鱼、家禽、酒类）。

购买经认证的鱼类和海鲜产品；不许供应由臭氧物质加工处理的食物。

从绿色农场购买有机食物和酒类；提供素食菜单。如卫生法规允许，可用剩余食物可提供给慈善机构；拒绝使用过度包装的食品。

选用可重复使用的容器运输食物；妥善储存、使用和处理危险清洁产品；确保员工经过适当的培训；寻找含有化学物质清洁剂的替代品；考虑与当地农业市场合作。

Chapter 2: Dining in Museums

2.1.2 Exhaust Design

Many museums are under-designed as it relates to visitor dining (and catering support). Limiting a café menu principally to cold menu items (salads and sandwiches) and items that can be prepared (warmed-up) with low-temperature cooking (soup, Panini style sandwiches and the like), will result in limiting the ability to provide the best possible visitor (and staff and volunteer) dining experience and optimise financial results.

If the museum wants to build or expand a Café or restaurant space to include onsite cooking and there isn't an option to install ducts and grease exhaust fans outside the building, there are other solutions people may want to consider.

Advanced exhausting equipments are installed in the open kitchen of Nerua Restaurant
Nerua餐厅内开放式厨房中安装了先进的排烟设备

One of the most significant trends in kitchen ventilation is going hoodless. Hoodless systems include self-contained ventless hood systems, ventless high temperature dishwashers, and separate ventless cooking equipment like speed ovens with built-in grease capture.

2.1.2 排烟设计

现如今，许多博物馆都开设了游客餐饮服务，但食物供应仅限的冷餐（沙拉和三明治）到低温加工食品（汤、意大利风帕尼尼三明治等），降低了为游客提供最佳就餐体验的可能性，从而减少了收入。

如想在博物馆内部打造或扩建包含现场烹饪的咖啡厅和餐厅空间，且不能在建筑外面安装管道和油烟排风扇，则可以考虑其他方式。

其中最为盛行的一种方式即为采用无烟罩式排风系统，包括无孔罩排烟系统、无孔高温洗碗机以及无孔分离烹调设备（内设油脂扑捉器的小型微波炉）。小型微波炉在目前备受欢迎，将对流技术与传统的微波能源相结合，规格和特性可以根据需要定制。附加的电热丝或陶瓷锅用于烘烤食物。

如需供应多种风格的食物，则无孔通风设备不失为一个合适的选择。无孔排烟罩带有内置油烟和气味过滤器，安装在电气烹饪设备（电热平扒炉、电炉和电炸锅）上方。此外，大多数设备设置灭火系统，减少火花喷射。但值得注意的是，这些设备要求日常清洁并需及时更换过滤器。

大多数排烟设备安装在墙壁或天花上，同时还有箱式移动设备，通过烹饪区底部或两侧将油烟排入设备底部的过滤结构。箱式设备被称作移动的艺术，可以移动到画室或接待区，供特殊活动时使用。

感应烹饪（食物通过不锈钢锅上面的磁场进行加热）可以有效地减少烹饪过程中的热量消耗，但却不具备吸收油烟和防火的功能。

Speed ovens have proliferated in the quick service market and they are available in a variety of sizes and optional features from companies. Most of these systems combine convection or impingement technology with microwave energy. Select models include browning elements or ceramic pans that allow them to be used for grilling.

When menus demand more variety, a ventless hood system may be a solution. Ventless hoods are installed over electric cooking appliances like electric griddles, ranges, and fryers and have built-in grease, smoke, and odor filters. Most systems have built-in fire suppression systems to reduce the likelihood of a flare-up creating a fire in the filters. These systems do require daily cleaning and replacing of the filters.

While most ventless systems are wall or ceiling mounted there are kiosk type systems where the air is drawn down the back or sides of the cooking area to a filter system under the equipment. Many are part of a mobile art that can be rolled into a gallery or reception area for special events.

It's important to note that induction cooking (where food is energised in a stainless pan using a magnetic field under the pan) is beneficial in reducing heat given off by the cooking process; it doesn't eliminate the need for capturing grease or providing fire protection from flair-ups on a pan or griddle.

Most ventless systems have little or no impact on reducing heat or odor generated by the cooking process. Ventless hoods or appliances used in non-traditional locations like hotel and office lobbies, atrium spaces, and small pantries where general ventilation is limited can produce odors if they are sucked into the general building exhaust. They also don't capture small particles of grease that can oat window, permeate walls and fabrics, or darken ceilings.

When planning a sit-down restaurant with a broad menu, a good choice is greaseless venting. These systems use technology developed for

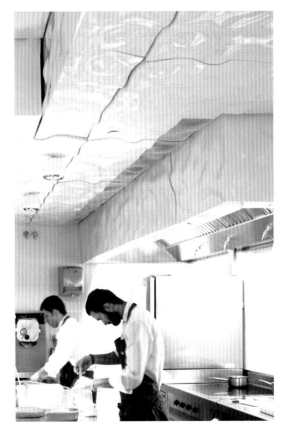

Advanced exhausting equipments are installed in the open kitchen of Nerua Restaurant
Nerua餐厅内开放式厨房中安装了先进的排烟设备

大多数无孔排烟设备在减少烹饪过程中的热量损耗和油烟产生没有多少影响，如放置在酒店、办公大厅、中庭以及小餐室内并与大厦的整体排烟系统相连，则会产生大量的气味。同时，这些设备不能吸收油脂颗粒，从而不可避免地使窗户、墙壁以及天花等被油渍侵蚀。

Chapter 2: Dining in Museums

use in nuclear submarines and combine a standard grease hood with a sophisticated pollution control unit that can remove 98-99% of odor, smoke, and grease. Most of these systems include built-in automatic cleaning of the electrostatic filters, so operating cost is reduced by not having replaceable filters. These advanced systems clean the air so that it can be exhausted directly into large volumes of space or through conventional sheet metal ductwork to the outside of the building.

When considering a ventless kitchen, heat generated from non-grease producing equipment like ice makers, toasters, sandwich grills, and dishwashers is an issue. There are specialty sandwich grills and dishwashers with built-in filters and collectors to reduce heat and steam. Heat and odor can often be reduced by installing general purpose fans and ductwork similar to those in bathrooms and showers that can be exhausted in louvres over doorways or through side wall louvres. Regardless of whatever system is installed, the exhaust for the food prep area should never be returned into the building make-up air. It needs to be exhausted separately and directly outside.

Like many products, ventless exhaust equipment can be misused and oversold, so investing in a dedicated exhaust system is likely to be a much better solution for the long term. Since only the most sophisticated (and expensive) systems can mediate odor, it's important to evaluate any impact the food aroma (or odor) will have on the patrons and staff.

如餐厅内需供应多种风格的食物，则除油脂通风设备是一个很好的选择。这些设备采用核潜艇生产技术，结合标准的油脂抽引罩结构和复杂的污染控制系统，可排除98%~99%的气味、油烟和油脂。这一设备安装了内置的静电清洁过滤器，从而削减了因更换过滤器而带来的成本预算。同时，其还可以净化空气——直接排除油烟等或通过传统的金属管道将其排向建筑室外。

使用无孔排烟通风设备时，制冰机、烤箱、三明治烤炉以及洗碗机等无油烟设备释放的热量处理也是应考虑的问题。带有内置热量和蒸汽过滤收集器的特制三明治烤箱和洗碗机设备、浴室中使用的排风扇（通过百叶窗等排除蒸汽）都可以被运用进来。值得注意的是，无论采用哪种排烟通风设备，都应确保食物准备区的废气不能重新回到建筑空气补给系统，两者需要独立安装。

同其他产品一样，无孔排烟设备也可能被不当或过度使用。因此从长久利益出发，选择适合的排烟设备显得更加重要。由于只有最为先进的设备才具备调节气味的功能，评估食物气味对于游客和员工的影响具有重要的作用。

2.2 Functional Design

2.2.1 Café Design

Concerning visitors' consumption, food purchasing is commonly regarded as a key purchase for those seeking a distinct, sensory experience.

Museums are venues where people tend to go on a non-everyday outing and their café spaces can be considered to serve as major 'pause' reflective 'places' – not just spaces – of meaningful significance during a visit. This suggests such Cafés are effectively an escape within an escape (a dream within a dream) within the museum visit experience, providing a reflective extension of the non-everyday museum/gallery visit for contemplation of the beautiful and the sublime as presented across the full visit. In this role, museum cafés should be designed to offer suitably integrated 'dreamscape' environments, synergistic to the museum/gallery core experience. The café may then even stand as a key motivator for attraction towards the pleasures of the museum itself.

2.2 功能设计

2.2.1 咖啡厅设计

关于消费方面,食物购买被称为是那些寻求独特感官体验的游客的主要购买行为。

博物馆通常是人们外出郊游的主要选择,而咖啡厅作为主要的"休憩"场所在参观过程中具有至关重要的作用。这就意味着"咖啡厅"是博物馆参观体验中的"休闲乐园",更是一个放松思索的空间。从这个角度来说,博物馆内咖啡厅应该营造出一种梦境般的氛围,与整体的参观经历相互呼应。更进一步说来,咖啡厅可以作为博物馆吸引游客的催化剂。

The Café in the Holburne Museum is a good place for the visitors to rest
赫尔本博物馆花园餐厅为游客提供了良好的休闲空间

Chapter 2: Dining in Museums

MOSI is a good example for Café design in museums, including good colours, good furnishing, and good layout
曼彻斯特科学工业博物馆咖啡厅和餐厅是博物馆内餐饮空间设计的范例，在颜色、装饰和布局上值得参考

① Integrated Planning

Good cafés were found to be a key attraction and motivation for museum visitation, with a visit to the café seen as an extension of the duration of a visit to the core exhibition spaces, providing 'a cognitive break' during the potential intellectual stresses or 'wear-out' of a visit's self-educational purpose.

Smart coffee bars are well established in bookshops, garden centres, leisure attractions and department stores. Museums have little choice – they have to consider providing what the customer wants; catering is now wherever people congregate and there is no better setting than a busy museum or one that has a good location.

Catering is a great business with healthy margins, lots of feel-good factor and a real capacity for creating a distinctive product. Quality food, drink and service can be a huge 'pull', rivalling the best of exhibits. There is no doubt a good café can dramatically increase museum footfall.

The range of catering options is large. Coffee kiosks, self-service cafés, assisted service coffee shops and full service restaurants are just some of the possibilities. Then people need to decide kitchen, no kitchen or the halfway house of a food preparation area with a simple contact grill. It is found that an assisted service café with a simple kitchen is what suits most small museums.

Museum café design should ideally encompass experiential signifiers to support visitors' contemplation of, or reflection upon, aspects of socio-cultural significance – the beautiful and the sublime – as represented by each distinct type of venue. The café space offers an extended immersion in time and space within a distinctly non-everyday, reflective ambiance. The design of museum cafés as spaces not only to eat in but also to dream in provides challenges to food service providers, but ones that could result in the cafés themselves becoming motivators for institutional visits and longer stays due to their particularly appealing and distinctive design dimensions.

① 整体规划

好的咖啡厅可以作为博物馆吸引游客的主要动力。游客可以在咖啡厅驻足休憩，从而延长在展区的参观时间。

书店、公园、休闲中心和百货商店内往往设计有精致的小咖啡吧，而博物馆选择性较少，并需要考虑顾客的需求。餐饮空间是人们休闲机会的最佳场所，而博物馆内不失为一个好的选址。

餐饮经营往往能够带来丰厚的利润，打造众多良好的因素并创造出独特的产品。优良的食品和优质的服务能够成为巨大的助力，一个好的咖啡厅可以大大地提升博物馆的客流量。

餐饮经营方式多种多样，包括小咖啡厅、自助咖啡屋、餐桌服务式咖啡店和全天候餐厅等。随后，便可考虑厨房的设计（即便没有厨房，也需预留备餐区的位置）。研究发现，餐桌服务式咖啡厅适用于多数小型博物馆。

博物馆咖啡厅设计应包含体验性的标识或符号，体现出社会文化内涵，便于游客思考。咖啡厅是博物馆整体环境中时间和空间的延续，不仅仅是一个就餐的场所，更是一个让人做梦的地方。当然，这一理念的提出为餐饮供应商带来了更大的挑战，但是成功的设计无疑能够成为提升博物馆客流的动力——独特而引人注目的设计吸引着游客长时间停留。

② Menu Selections

The menu is an all-important first step. Professional planning of the servery, kitchen, store and equipment can only start when the scope of the menu is established.

The beauty of a traditional café is that it can be designed to be customer friendly and cost effective in terms of labour costs and skills. Self-service of as many items as possible is the key.

Cold savoury food revolves around filled sandwiches and baguettes. Life starts to get more challenging with the provision of hot food. Soup is an easy win and jacket potatoes and toasted sandwiches/panini are popular. Hot food brings up the question of whether a 'customer waiting' system (using numbers, flags, wooden spoons or whatever) is needed to avoid the queue becoming unmanageable.

More complicated hot food may take the Café into the realm of needing cooks, more complicated equipment, extract ventilation and more complexity. The food menu will allow people to specify the correct amount of equipment in the least amount of space and enable the utility requirements to be correctly sized.

Cakes and traybakes are central to most cafés. Customers buy with their eyes, and good display (just as with your exhibits) is vital. Cream cakes will need refrigerated display but don't chill the remainder of the cakes to death – the starch in cakes appreciates ambient/room temperature storage. Home-produced scones, particularly those baked on a daily basis, are highly popular and profitable. Local suppliers of cakes can be a real plus but do carry out and record a risk assessment of how they are produced and stored.

Protective film (but never cling film and preferably not those cloudy plastic domes) will of course be required to protect the food, but try to ensure presentation is as innovative, imaginative and appealing as the displays in the rest of the museum.

The menu of Barbican Lounge is redesigned to service the visitors in a better way
巴比肯艺术中心休闲餐厅内的菜单重新设计，便于更好地为游客服务

② 菜单选择

菜单选择是重要的起点——菜单确定之后，备餐室、厨房、店面和设备的规划设计才可以实施。

传统风格咖啡厅的吸引力在于其温馨友好的顾客服务以及经济有效的经营方式。自助式服务以及多种类型的食物供应往往起到至关重要的作用。

咖啡厅起初只供应三明治和法式长棍面包等冷餐食物，随后开始提供热餐食品，当然这也带来了更多的挑战。汤品轻松获得顾客的亲睐，英式烤土豆和烘烤三明治备受欢迎。热餐食物的供应同时带来了一个问题——是否应该设定"顾客等候"机制（采用序号、标识、木板等区分等排序），以避免出现队列无法控制的局面。

Chapter 2: Dining in Museums

The food provided in Barbican Foodhall is various and fresh
巴比肯艺术中心休闲餐厅供应的食物种类多样，美味新鲜

Most catering outlets now have quality espresso coffee equipment that avoids the powder and concentrate systems of old.

The cold drink menu will allow the correct sizing of a refrigerated self-service display cabinet. Freshly squeezed orange juice is a winner although just beware of the cleaning implications of the 'squeeze-on-demand' machines. We tend to recommend bottled soft drinks rather than the great margin, post mix/syrup system on the grounds of better quality and flexibility.

Above all, exercise discipline with the menu. It should not indulge a chef's fancy or be salesman-led. It is not a retail shop where extensive choice is good – a vast range of similar products leads to customer confusion and slow café counter service speeds. The menu can stay the same for extensive periods as it is rare for a museum to have repeat visitors on a weekly basis. Menu discipline should lead to the tasting of all products prior to their appearing on the menu, greater consistency and higher quality.

咖啡厅需考虑雇用大厨、引进更复杂的设备、打造通风结构，这些都是提供热餐食物所必须配备的。咖啡厅可以根据菜单正确估算设备的数量，在有限的空间内实现最佳的使用方式。

蛋糕类食品是大多数咖啡厅供应的主要食物。良好的视觉吸引力往往能够促使顾客进行购买，因此陈列展示显得格外重要。奶油蛋糕需要冷藏陈列，但同时需确保其他部分的适宜温度（如淀粉热衷于常温环境）。家庭自制的烤饼（尤其是每日现烤的）备受欢迎，并能够带来更多的利润。选择当地蛋糕生产商供应的食物不失为一个好的选择，但要仔细评估其制作过程和储藏方式。

保护膜（不是保鲜膜或者硬塑料材质）用于保护食物，但同时确保其陈列方式具备创新性、想象力和吸引力。

多数餐饮空间拥有浓缩咖啡设备，避免粉末喷溅等。冷饮菜单决定着自助式冷藏柜的规格和陈列方式。鲜榨橙汁备受亲睐，但要注意榨汁机的清洗问题。我们推荐供应瓶装饮料，除了利润因素之外，还能保证质量。

总之，菜单的制定需慎重考虑，既不能完全根据大厨的喜好，也不能任凭推销员的指导。明确这里并不是一个零售商店，并不是选择越多越好——过多的类似产品容易使顾客迷惑，减慢服务速度。很少有人会重复地参观同一个博物馆，因此其菜单可以在长时期内不必更换。菜单上的食品在供应之前，应进行评估，确保连续性和高质量性。

第二章　博物馆中的餐饮设计

The terrace dining area in Barbican Centre
巴比肯艺术中心休闲餐厅露台就餐区

Beverage service in Coach House Restaurant
哈特菲尔德庄园餐厅内饮品服务区

③ Space Planning

One hundred square metres is about the smallest space requirement for a servery/open kitchen, store and 40-person seating area. Electricity, drinking water, hot water, drainage and ventilation are essential.

The caterer will always want a highly visible, accessible location near the front of the building to maximise sales. Natural light, some sort of view and summer outside seating all make a substantial difference. As with the museum itself, catering is about location.

④ Equipment Design

Common mistakes at this stage include having just one long counter in a busy café (queues build up for the hot food and those just wanting a coffee and cake have to tediously wait in line) and including items such as salad bars that have notoriously slow self-service times. Good design can feature a number of separate food counters, a large element of self-service (including hot drinks) and really appetising presentation.

③ 空间规划

咖啡厅规格不应少于100平方米，包括备餐室/开放式厨房、存储区和供40人使用的就餐区。确保电气、饮用水、热水的供应，排水和通风系统保持良好。

餐饮服务商往往热衷于可见性较强的位置——建筑入口附近，以提高销售利润。自然光线、别致的风景以及室外就餐区等可以使其看起来更加与众不同。博物馆内的餐饮空间，选址才是关键。

④ 设备设计

最常见的失误是在一个繁忙的咖啡厅内只设置一个长柜台，造成只需一杯咖啡和一块蛋糕的顾客要同需要热餐的顾客一同排队等候。好的设计需要设置多个独立的柜台和自助服务项目，同时确保良好的展示方式。

Chapter 2: Dining in Museums

People should be aware of the trend to 'contact grill kitchens' where this one humble piece of equipment is all that is used to cook a wide variety of panini, breakfast and short order grill items. The full kitchen fit-out of old is fast becoming an unnecessary expense.

Two excellent modern items of equipment are the small bake-off oven located on the servery back counter (for that tantalising aroma of fresh croissant) and the expensive, but ultra-fast turbochef oven.

⑤ Display Design
The display skills so evident in most modern museums can be used to excellent effect in the café.

单压板扒炉成为厨房中的重要的烹饪设备，可以制作帕尼尼（意大利三明治）、早餐以及其他烘烤类食物，完全可以取代厨房中的传统烹饪设施。

现代化的烹饪设备包括放置在备餐区后侧柜台上的小型烤箱和价格昂贵的快速微波烤箱。

⑤ 陈列设计
多数现代博物馆中都以恰到好处的陈列技术而著称，当然这一技术也可用于咖啡厅设计中。

The display designs in MOSI
曼彻斯特科学工业博物馆咖啡厅和餐厅陈列设计

⑥ Price Decisions

The designers can't do anything about the pricing of the food, but they hear concerns about food service prices all the time, especially from families. While it might seem logical that the food service would be a profit centre, this goal is rarely achieved. The experience is that food service is an amenity provided to help people to extend their visits. If visitors find the food to be a good value and the venue convenient and comfortable, then the food service is successful, even if it only breaks even financially. Of course, there is money to be made with special events and catering. These too serve the mission by bringing new people to the museum, but they can also generate significant revenue. Planning spaces for the caterers is as important as planning for the visitor's food service needs.

⑥ 价格设定

关于食物的定价，设计师不能给出任何意见，但关于餐饮服务空间的价格他们却耳熟能详。餐饮服务空间往往被认为是一个利润中心，但实际上这一目标很难达到。餐饮空间可以在一定程度上延长游客参观的时间——如果游客对于餐饮空间的评价很高（食物优良，环境舒适），那么其便是成功的，即使并未带来很大的经济效益。当然，餐饮空间可以通过举办特殊活动增加收入，同时可以为博物馆吸引更多的游客。因此，在博物馆中打造餐饮空间同满足游客的就餐需求一样重要。

The price tag of MOSI
曼彻斯特科学工业博物馆咖啡厅和餐厅内食物价签

Chapter 2: Dining in Museums

2.2.2 Restaurant Design

① Circulation Diagram

Museums need at least two circulation schemes, one for the public and another for the back of house functions. The food should be able to find its way from the loading dock to the kitchen and to special events areas without crossing through any public space. Museums with high value collections (art museums, generally) need a third circulation system for works of art. Art deliveries should have their own loading area, physically separate from the area where food is delivered.

② Location Selection

If restaurant is an important part of the visitor experience, why are so many venues in the back of the basement? Of course, exhibit spaces come first, but the restaurant also needs to be easy to find. Ideally, the restaurant area will be right off a circulation hub in the free zone, although not necessarily at the entrance or off the main lobby. Visitors should be able to see outside or into a lobby or other gathering space, allowing them to feel like they are still part of the museum experience while they eat. Of course, in a busy science centre, it may be necessary to have a more isolated area for school groups – but even then, the space should be easy to find.

③ Seating Capacity

Planning for restaurant capacity is as much art as science. Too often, there

2.2.2 餐厅设计

① 动线规划

一个博物馆至少需要两条动线,一条为公众所用,一条为博物馆内部所用。食物运输路线不应经过任何公共空间,应从卸载区直接到达厨房和举办特殊活动的区域。具有较高收藏价值的博物馆(如艺术博物馆)还应具备第三条动线,用于艺术作品的传送。艺术品应具备单独的卸载区域,与食品传送完全分离开来。

② 选址

如果餐厅被视作游客体验的重要组成部分,为何大部分被选址在地下室内?展览空间固然重要,但餐厅选址也应遵循可见性强的原则。虽然不必设置在入口或大厅附近,但最好选择空间动线枢纽区域内,确保游客在餐厅内可以看到室外风景或者大厅等公共空间的景象,使就餐过程成为博物馆体验的一部分。当然,在游客众多的科学中心内,往往根据流派分布多个独立的区域。即便如此,餐厅也应选址在显而易见的位置。

The Barbican Foodhall is a good example of location selection
巴比肯艺术中心休闲餐厅在选址上格外成功

第二章　博物馆中的餐饮设计

just isn't enough space – especially if the budget is tight. Most museums don't need 1,000 seats, but having enough seats will allow visitors a chance to relax, rest up, and spend more time in the exhibits. The time dedicated to recharging results in greater visitor satisfaction. On the other hand, too many seats can make the space feel deserted. Unfortunately, there are no rules of thumb. The best bet is to work with someone who is experienced in planning and designing seating areas for similarly scaled facilities with similar audience profiles. This is another area where a little extra space can make a big difference.

③ 座位容量

餐厅座位数量的规划也是一门科学艺术。多数情况下，由于预算有限，餐厅规格往往较小。一方面，多数博物馆并不需要1000个就餐座位，但充足的座位可以确保游客更加舒适地休息，从而花费更多的时间参观。通常，休息时间的长短决定着游客的满意度。另一方面，过多的座位会使空间闲置。不幸的是，关于餐厅座位数量设定并没有规章可循，最好的方式就是同有相关经验的设计人员共同合作。这就意味着细节有时可以起到决定作用。

Seating area is well organised in Coach House restaurant, and Nerua Restaurant
在哈特菲尔德庄园餐厅和Nerua餐厅内，座区设计格外突出

Chapter 2: Dining in Museums

Seating area is well organised in Barbican Lounge
在巴比肯艺术中心休闲餐厅内,座区设计格外突出

④ Considering the Queue
Managing the circulation and traffic in a busy self-service or Caféteria-style food service area can be a challenge, especially when space is already cramped. On busy days, bottle-necks may be inevitable, but attention to the design of the queuing area can alleviate most of the everyday congestion so that people can easily get what they want and staff can serve them efficiently. The help of experienced professionals during planning and design is the best way to avoid queuing problems.

④ 排队问题
在自助式餐厅内,实现有效的交通流动管理是一个很大的挑战,尤其是在一个相对拥挤的空间内。在繁忙的营业时间内,不可避免出现问题,但对排队区域的有效设计可以避免拥堵状况,从而让顾客方便地拿到自己的食物。在规划和设计过程中向经验丰富的专业人士咨询可以很好地处理这个问题。

⑤ Kitchen Design
A good cook can make do in any kitchen – they've all had a phenomenal meal made in an impossibly small kitchen – but no one wants to work like that every day. The right kitchen begins by determining the properly sized space needed to prepare the type and number of meals the museum will need. The specific layout then needs to be worked out by a professional kitchen designer based on the specific needs of the type of food service to be provided. That design needs to be integrated into the overall design to allow for adequate power, and access to water, waste, and gas lines, as well as adding appropriate ventilation. Lack of venting for cooking areas can be a real problem if it isn't planned for in advance. Kitchen design is an area where people often feel like instant experts – but just because the restaurant manager has been cooking for years doesn't mean he or she knows how to design a kitchen.

⑥ Food Storage
One of the biggest complaints people hear is that there is not enough space to store bulk food items. Sometimes there is plenty of cold storage, and no dry storage – or vice versa. Plan for a little extra storage and the food service staff will find a good use for it.

⑦ Waste Disposal
It seems obvious that a restaurant produces trash, but less obvious is that restaurant waste includes a high percentage of food. Moreover, if it is not carefully handled, food waste can get smelly, attracting pests and vermin. Be sure to provide a separate, secure space for food waste – and make sure the containers are emptied regularly and that the area is kept spotless.

⑤ 厨房设计
一个好的厨师可以在厨房中做出任何美味的食物，即便是在一个格外狭小的厨房中也能做出美味大餐，但没有任何人希望终日在拥挤的空间中工作。首先，一个合适的厨房在规格上应与博物馆所需要的食物类型和食物数量相互呼应。其次，专业的厨房设计师根据需求规划布局。最后，厨房设计应考虑整体因素——确保足够的水、电、气供应以及有效的垃圾处理和通风方式。在规划前期没有充分考虑通风将是一个很实际的问题。在厨房设计领域，人人都觉得自己是专家。但实际上，即使是一个拥有多年烹饪经验的餐厅经理有时也并不清楚如何设计厨房。

⑥ 食物储存
经常听到人们抱怨没有足够的空间储存散装食品。通常情况下，餐厅内配备足够的冷库空间，但没有干存储空间或者存在相反的情况。

⑦ 垃圾处理
显而易见，餐厅会产生垃圾。其大部分垃圾是食物，如处理不当就会散发气味，招致害虫等。因此，需设立单独而安全的空间用于存放食物垃圾，同时确保盛放容器定期清空，空间一尘不染。

Case Studies
案例赏析

West Valley Art Museum Café Renovation
西部峡谷美术馆咖啡厅翻新

Untitled
未名咖啡厅

Nerua Restaurant
Nerua 餐厅

The Wright
赖特餐厅

Holburne Garden Café
赫尔本博物馆花园咖啡厅

Groninger Museum Restaurant "Mendini"
格罗宁根博物馆 "门迪尼" 餐厅

MOSI
曼彻斯特科学工业博物馆咖啡厅和餐厅

Coach House, Hatfield House
门房改造餐厅

L'Osteria Künstlerhaus
艺术之家活动中心 L'Osteria 意式餐厅

The Whitechapel
白教堂画廊餐厅

West Valley Art Museum Café Renovation

西部峡谷美术馆咖啡厅翻新

Completion date: November 2007
Location: West Valley Art Museum, Surprise, Arizona, USA
Designer: colab studio, llc
Structural: JT Engineering, Inc.
Electrical: Woodward Engineering, Inc.
Photographer: Bill Timmerman, Timmerman Photographer Inc.
Area: 204m² (2,200 SF)
Main materials: Birch flooring, birch cabinetry, fabric, tile, stainless steel, painted steel, vinyl graphics

完成时间：2007年
地点：美国 亚利桑那 瑟普赖斯 西部峡谷美术馆内
设计： colab工作室
结构设计：JT工程公司
电气设计：Woodward工程公司
摄影：比尔·蒂默曼/蒂默曼摄影公司
面积：204平方米
主要材质：桦木地板、桦木橱柜、织物、瓷砖、不锈钢、涂漆钢板、乙烯图案

The West Valley Art Museum is located in Surprise, Arizona, which was formed as an 'active retirement community' on the outskirts of Phoenix. The museum was greatly desired during its formation in the 1970s, and grew in popularity and size in the 1980s. Funding in the 2000s began to shrink, however, as the population aged, and the museum's membership declined.

To make matters worse, the greatly out-of-date café was costing more money to the museum than it was bringing in. The new general manager of the museum decided to get rid of the company running the café, and refurbish the space in an effort to attract younger people into the space and entice them to become members at the museum.

1. New exterior entry
2. Exterior terrace
3. Café dining
4. Bar / preparation area
5. Kitchen
6. Male WC
7. Female WC

1. 新外部入口
2. 外部露台
3. 就餐区
4. 吧台 / 备餐区
5. 厨房
6. 男卫生间
7. 女卫生间

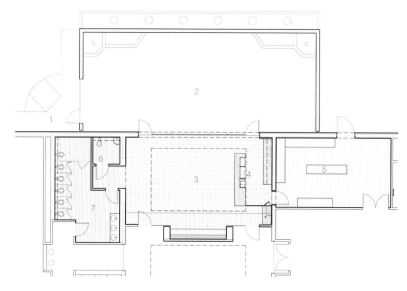

floor plan of café renovation 咖啡厅翻新平面图

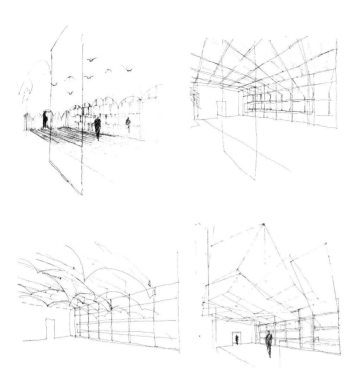

However, it was important the new café design would not impede the museum's activities or functions, and not alienate the elderly patrons, whose average age is about 85. To the designers, this was the most difficult task: to attract a younger, more design-sophisticated crowd, and still appeal to a fairly conservative community of aging art appreciators.

The space itself was also a great challenge. One side is all glass facing a very poor view of low-cost housing beyond the courtyard. The other long side of the space faced the museum, which required a maximum of wall space. The ceiling was a bit too high in proportion for the room, which was very small to house all the seating desired. A new service bar needed to be added, as well as a new entry to men's and women's restrooms.

The budget was amazingly small for such a project, and limited many of decisions to finish materials. So, the design started from the idea of making a space that felt enclosed, warm, and comforting. The designers decided to use wood for the floor and millwork to provide a residential feel in a commercial setting. They also elected to utilise a reoccurring theme of residential roof forms at a 15-degree pitch, to feel familiar to people in the community. The angle was chosen to reflect the dominant roof type of the community, mostly built in the 1960s and 70s.

The design is meant to be an all birch wood 'base' of floor and millwork sitting beneath translucent cloth 'roofs' that bring the scale down some, but still leave the room feeling spacious. They broke down the institutionally commercial glazing system with a series of angled graphics, allowing light into the

Evening shot of café from entrance　咖啡厅入口夜景

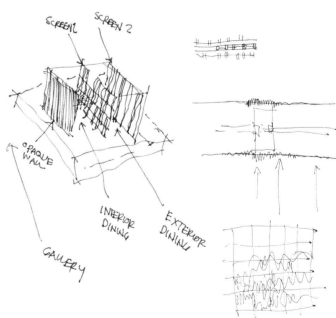

Preliminary design sketch of opacity sequence　手绘图

Exterior view looking into dining area 从室外看就餐区景象

space, but obscuring the view. The bar echoes the angled forms with custom-built wine and glass racks all built of birch wood. The restrooms are clad in a rich, dark tile with birch wood partitions. The forms, materials, and details are simple, but allow a rich, vibrant, and elegant space to function beautifully.

西部峡谷美术馆位于亚利桑那州瑟普赖斯地区，这里构成了凤凰城郊区的"退休人员社区"。20世纪70年代，博物馆备受欢迎，这种状况一直持续到20世纪80年代。21世纪开始，美术馆基金开始缩水，该地区的人口步入老龄化，从而导致美术馆会员逐渐减少。更糟的情形是，过时的咖啡厅面临着产出低于投入的困境。至此，美术馆负责人决定重新经营咖啡厅，并对其进行翻修，旨在吸引更多的年轻游客，同时将他们发展成美术馆会员。

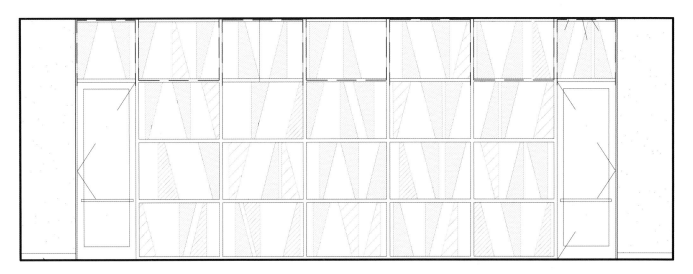

Elevation of exterior glass vinyl pattern 带乙烯图案玻璃结构外立面

Dining area before renovation 翻新前就餐区

Dining area after renovation 翻新后就餐区

新咖啡厅的设计不能影响到美术馆的正常运营,并且不能忽略平均年龄 85 岁的老年游客,这一点至关重要。对于设计师来说,这更是一项极具难度的任务——既要吸引年轻一族,又要满足老年一代的艺术鉴赏水准。

空间本身也带来极大的挑战——一面是全玻璃空间,将对面低成本住宅区的贫穷景象映射进来;另外一面朝向美术馆,需要最大限度打造墙壁结构;空间过于狭小,不能容纳所需的就餐区,而天花对于空间规格来说过于高挑。除此之外,要求增添一个新的服务柜台和通往男、女卫生间的入口。

较少的预算限制了装饰材质的选择。设计师决定打造一个封闭、温馨而舒适的空间,采用木质地板和木艺家具营造居家氛围。天花设计模仿当地社区民居样式,使其倾斜 15 度。这是这一地区建于 20 世纪 60、70 年代民居的一大特色。

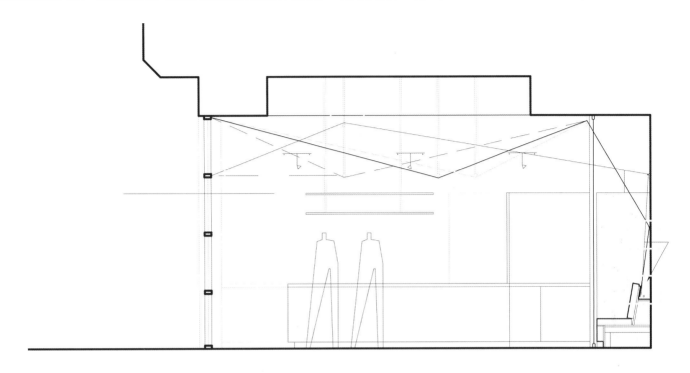

Interior elevation of dining area　就餐区室内立面

Evening shot of Café interior　咖啡厅室内夜景

View through wine bar glassware 酒吧区玻璃器皿

设计中大量运用了桦木材质，如桦木地板和半透明屋顶下的木艺装饰（在视觉上降低了天花的高度，但却不影响空间的开阔感）。全玻璃一面采用角状乙烯图案装饰，既不影响光线透射，又能遮挡室外贫穷的景象。吧台上的酒架和杯架同样选用桦木打造，角状图案反射到吧台上，别具特色。卫生间采用深色瓷砖装饰，并使用桦木板间隔。图案、材质和细节装饰虽然简单，但却营造出一个充满活力的典雅空间。

Interior elevation of bar 吧台室内立面

Preparation area near entry 临近入口的备餐区

Restroom before renovation　翻新前卫生间

View into restroom　翻新后卫生间

Graphics on women's restroom door　女士卫生间门上图案

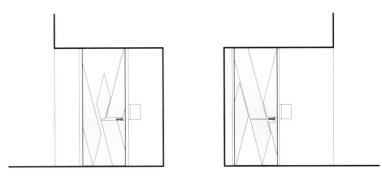

Interior elevation of vinyl graphics on restroom doors　卫生间门上乙烯图案室内立面

Untitled

未名咖啡厅

Completion date: April 2011
Location: Whitney Museum of American Art, New York, USA
Designer: Rockwell Group
Photograper: Blandon Belushin, Paul Warchol
Area: 139.35m²
Seating: Dining area 70–84, Bar 10
Client: Whitney Museum of American Art, Union Square Hospitality Group

完成时间：2011年
地点：美国 纽约 惠特尼美术馆内
设计：罗克韦尔集团
摄影：布兰东·拜鲁辛、保罗·瓦尔霍乌
面积：139.35平方米
座位：就餐区（70~84）、酒吧区（10）
业主：惠特尼美术馆、联合广场酒店集团

Rockwell Group has joined forces once again with famed restaurateur, Danny Meyer to design a new restaurant and café at The Whitney Museum of American Art. The restaurant, opening in April 2011, will be operated by Mr. Meyer's Union Square Hospitality Group.

罗克韦尔集团强强联手，再次为著名的餐馆老板丹尼·迈耶设计一个新的餐厅和咖啡馆，位于惠特尼美国艺术博物馆。这个餐厅将由迈耶先生的联合广场餐饮集团运作，于2011年4月开始营业。

To complement the mid-century modern architecture of the Marcel Breuer building, the cellar-level restaurant offers an uptown take on the traditional diner with streamlined furniture and features.

Rockwell Group has designed simple white oak tables and room dividers, seating with red felt upholstery, and custom metal lamps to create a comfortable and industrial-chic environment.

Felt-backed white oak benches line three sides of the restaurant with long communal tables filling the interior. All furniture and fixtures can easily be removed for museum events.

为完善本世纪中叶马塞尔·布鲁尔的现代建筑风格，流线型的家具和特色设计为这个位于地下室的传统风格餐厅增添了现代气息。

罗克韦尔集团设计了简单的白橡木桌子和房间隔断，红色毛毡装饰的座椅和定制的金属灯具，创造一个舒适而别致的工业化氛围。

背部装饰着毛毡的白色橡木长凳围合在周围，与长桌一起填满了整个空间。所有的家具和装置都可随时拆卸，便于美术馆举办各种活动。

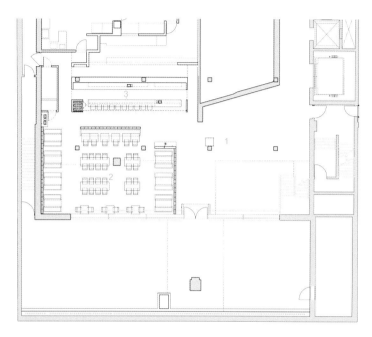

1. Waiting area
2. Main dining space
3. Café / bar

1. 等候区
2. 主用餐区
3. 咖啡 / 吧台

Cellar level　地下室层

Delineated by a large white lacquer frame, the bar area is backed by a chalkboard detailing daily specials and featuring a wood cake display shelves.

The polished white oak bar continues the use of wood throughout the restaurant while adding some modern touches: a Corian panel below and chrome and white leather stools. Museum-goers can visit the bar for wine and liquor, coffee and dessert.

吧台区被框入一个白色大油漆框架之中,后方的黑板上面详细地记录着每日的特价菜并摆放着一个木制蛋糕陈列架。

抛光白橡木吧台延续了餐厅的木质风格,底部的可耐力板和铬合金白色皮凳带来了一丝现代气息。游客可以在这里享用葡萄酒、咖啡和甜点。

Rockwell Group decided to keep the original stone floors and concrete architecture of the building intact to preserve the authenticity of the space.

Chrome-plated light bulbs line the ceiling of the restaurant to echo the iconic lamps designed by Breuer for the museum's lobby.

罗克韦尔集团保留了原来的石质地板和混凝土建筑主体，以凸显空间的真实性。

餐厅天花板上成排的镀铬灯饰与美术馆前厅布鲁尔设计的标志性灯具相呼应。

Nerua Restaurant

Nerua 餐厅

Completion date: 2011
Location: Museo Guggenheim, Bilbao, Spain
Designer: ACXT Architects
Photographer: Aitor Ortiz

完成时间：2011年
地点：西班牙 毕尔巴赫 古根海姆博物馆内
设计：ACXT建筑师事务所
摄影：埃托尔·奥尔蒂斯

Nerua Restaurant is located in Museo Guggenheim, Bilbao and gets its first Michelin Star just six months after its inauguration.

餐厅位于毕尔巴赫古根海姆博物馆内，在开业短短半年时间变荣获了米其林一星殊荣。

The project involves the restructuring of the museum Caféteria, exclusive to visitors, and its surroundings (hall, washrooms and preparation area) into a high standing restaurant, open also to the public and with a capacity for 40 diners.

这一项目是将原博物馆食堂（曾经只对游客开放）以及周围空间（大厅、洗手间、备餐区）等改造成餐厅，可容纳40人就餐，面向公众开放。

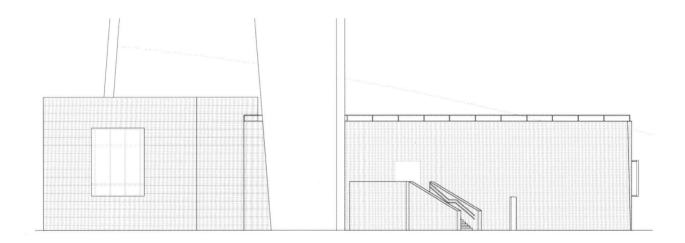

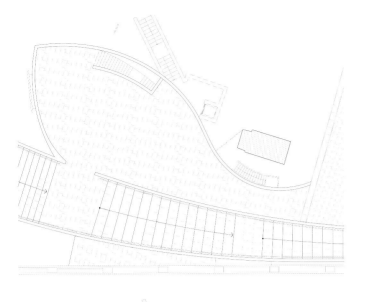

1. Facilities   1. 设备室
2. Storehouse   2. 储藏室
3. Distributor 3. 配送处
4. Female WC   4. 女卫生间

Terrace plan 露台平面图

1. Dining room 1. 用餐区
2. Lobby   2. 前厅
3. Clean   3. 清扫室
4. Storehouse   4. 储藏室
5. Office   5. 办公室
6. Wash   6. 洗碗间
7. Kitchen   7. 厨房
8. Male WC   8. 男卫生间

Dining room plan 用餐区平面图

The work was taken on with a clear objective: the space had to represent chef Josean Alija's culinary style; it had to be a true reflection of his creative processes.

设计的目标非常明确：其一即为通过空间展现名厨 Josean Alija 的烹饪风格；其二即为真实呈现其创意烹饪过程。

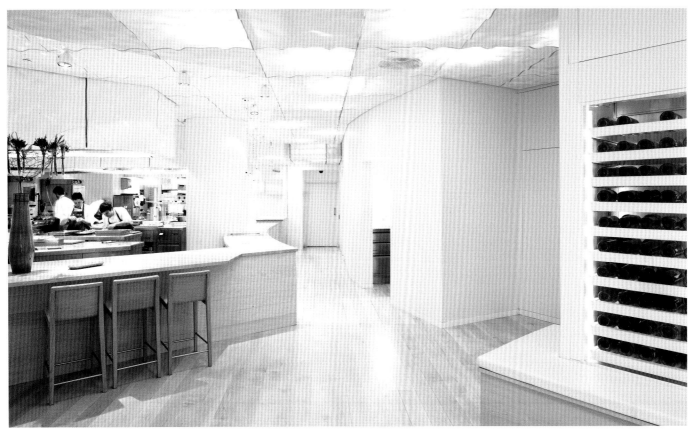

Maple wood and lacquered sheet are also shared, the latter being pierced in the restaurant to allow light, air conditioning and sound through and smooth and plain in the kitchen and the extractors.

The generated ambiance transmits simplicity, excellence and elegance.

枫木和喷漆金属片是主要材质。喷漆金属片可允许光线、空气和声音穿过，主要用在厨房内。

空间散发着简约而高雅的气息。

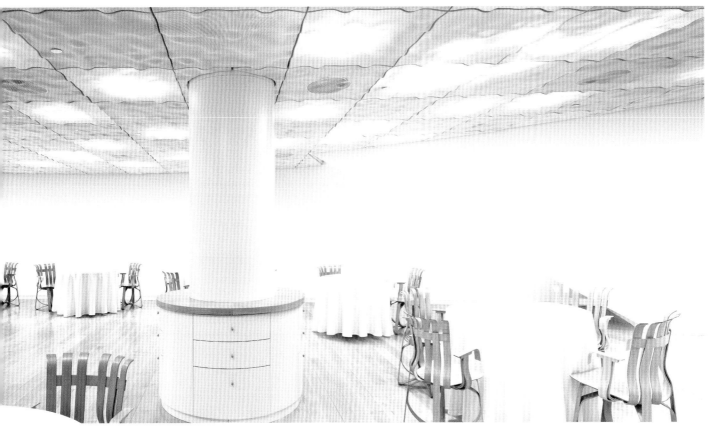

The Wright

赖特餐厅

Completion date: 2009
Location: Guggenheim Museum, New York, USA
Design: Andre Kikoski Architect
Photographer: Peter Aaron

完成时间：2009年
地点：美国 纽约 古根海姆美术馆内
设计：Andre Kikoski建筑事务所
摄影：彼得·艾伦

Though it's approaching the end of its year-long 50th anniversary celebrations, New York's Solomon R. Guggenheim Museum is not through with the festivities yet: December 11 saw the grand opening of its newest restaurant, The Wright.

Named after the Fifth Avenue museum's architect, Frank Lloyd Wright, the 58-seat restaurant was designed by Andre Kikoski Architect and features a modern American menu by David Bouley protégé Rodolfo Contreras featuring local seasonal dishes.

"It was both an incredible honour and an exhilarating challenge to work within Wright's iconic building," said Kikoski. "Every time we visit, we see a new subtlety in it that deepens our appreciation of its sophistication. We sought to create a work that is both contemporary and complementary."

"Inspired by and created within an institution renowned for its art, architecture and innovation, The Wright will extend that experience to its food and service," adds the restaurant's director, Aaron Breitman. "The Wright will appeal to neighbours seeking stylish and sophisticated dining as well as visitors who want to experience the thrill of New York in one of the city's greatest cultural treasures."

位于纽约的所罗门 R. 古根海姆美术馆虽然即将结束为期一年的 50 周年庆典活动，但其喜悦接踵而至——2009 年 12 月 11 日其全新的餐厅"赖特餐厅"（The Wright）举办了盛大的开幕。

餐厅以第五大道博物馆的建筑师弗兰克·劳埃德·赖特（Frank Lloyd Wright）的名字命名，共有 58 个座位，并由建筑师 Andre Kikoski 设计。餐厅以现代美国菜为特色，由名厨大卫·博雷（David Bouley）的接班人鲁道夫·孔特雷拉斯（Rodolfo Contreras）主理。

"在赖特的标志性建筑里面工作是一个令人难以置信的荣誉，同时更面临着令人振奋的挑战"，Kikoski 认为，"在每次来到这里，我们都能看到它全新的微妙之处，更能进一步加深我们对成熟干练的理解。我们寻求打造一个空间，既有时代性又有补充性。"

"在这样一个以艺术、建筑和创新著称的建筑里，我们备受鼓舞和启发，赖特餐厅将会把体验延伸到食物和服务中去。"餐厅的董事亚伦·布雷特曼（Aaron Breitman）接着说到，"赖特餐厅将吸引那些追求时尚和精美饮食的食客们，以及那些想体验纽约这一最大的文化瑰宝将带给他们怎样的激动的游客们。"

Centerd on a communal table, Kikoski's channeled Lloyd Wright's iconic spiralling white design from 1959 in the nearly all-white 1,600-square-foot space.

餐厅位于1959年面积约147平方米的近乎纯白空间内，设计以共享桌为中心，把餐厅打造成螺旋形状。

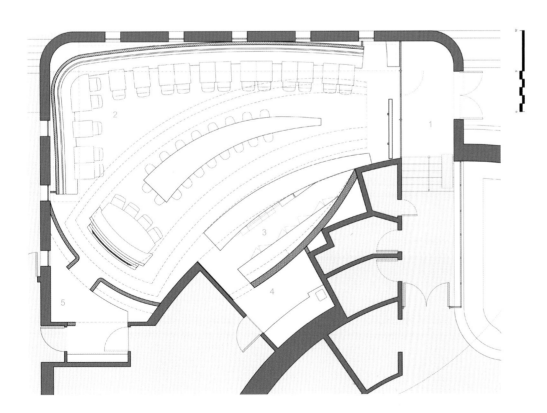

1. Vestibule
2. Restaurant
3. Bar
4. Back bar service
5. Service

1. 门厅
2. 餐厅
3. 酒吧区
4. 酒吧后服务区
5. 服务区

Floor plan　平面图

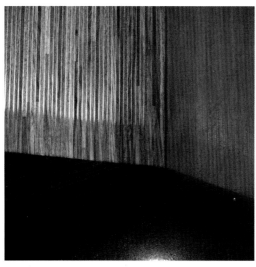

Standouts include a curvilinear walnut wall layered with illuminated fiber-optics and a bar topped in seamless white Corian and clad in a shimmering skin of custom metalwork.

引人瞩目的特色结构包括曲线核桃木墙，照明的光纤经过无缝的可丽耐面材，给金属制品镀上了一层亮眼的闪光。

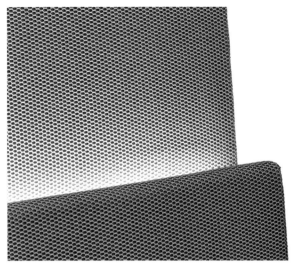

A sweeping banquette is glammed up with vivid blue leather seating backed by illuminated planes of woven grey fabric underneath a layered ceiling canopy of taut white fabric.

The ceiling is also home to the restaurant's signature artwork, a colourful site-specific sculpture of horizontal powder-coated aluminium planks by British artist Liam Gillick that marks the entrance and descends from overhead to cover the Wright's walls – just like in the museum's galleries.

干净大方的沿墙软座，装饰着蓝色皮革坐垫，天花板在光线的辉映下，与秩序井然的白色织物相得益彰。

天花板也是餐厅签名艺术作品的所在地，其中多彩的水平定位铝板雕刻是英国艺术家利亚姆·吉利克的杰作，它从门口处开始蔓延，从上到下覆盖了赖特餐厅的墙体，使得顾客犹如置身在博物馆的画廊中。

Holburne Garden Café
赫尔本博物馆花园咖啡厅

Completion date: May 2011
Location: The Holburne Museum, Bath, UK
Designer: Softroom
Photographer: Richard Lewisohn
Client: Benugo

完成时间：2011年
地点：英国 巴斯 赫尔本博物馆内
设计：Softroom建筑事务所
摄影：理查德·莱文森
客户：Benugo

The Holburne, a museum housed in a Grade 1 listed building at the entrance to the Sydney Pleasure Gardens in Bath, was recently restored and wonderfully extended by Eric Parry Architects. The Garden Café is situated at the heart of the scheme, in the ground floor of the new extension, where Benugo asked Softroom to create a café for around 55 people, including a servery where food could be prepared.

赫尔本博物馆位于悉尼游乐场花园入口处的一幢国家一级保护建筑内，其修复与扩建工程由埃里克·帕里建筑事务所（Eric Parry Architects）负责。花园餐厅位于博物馆扩建部分的一层，Softroom建筑事务所受托打造这一空间，包括备餐区，可容纳55人就餐。

Being largely glazed meant that the space could have felt exposed and uncomfortable in the winter months, so the recessed lighting was replaced with glowing pendants over the tables to add warmth and focus.

玻璃外观意味着在冬季的几个月时间内让人感觉不太舒适,因此设计师采用吊灯代替内嵌的灯饰,增添温暖的气息。

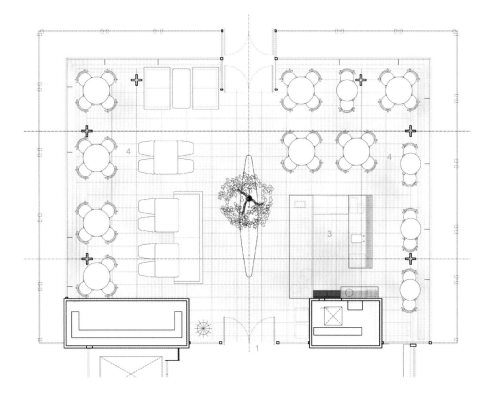

1. Entrance
2. Central display
3. Service counter
4. Dining area

1. 入口
2. 中心陈列区
3. 服务柜台
4. 就餐区

Floor plan 平面图

A central island display helps link together the two halves of the room, which could have felt too separated by the route to the garden running through the middle of the space.

穿过餐厅中央并通往花园的小路似乎将空间割裂，设计师专门打造了位于中心区域的陈列柜，在视觉上两部分空间结合起来，弥补这一缺陷。

The café has been an overwhelming success. The painted ceiling adds intimacy and helps link the space visually with the ceramic cladding of the building above and the greenery of the gardens all around.

这一咖啡厅的设计取得了巨大的成功。彩绘天花增添了亲切感，同时将空间在视觉上与建筑瓷砖覆层和室外绿意盎然的景色统一起来。

Other details, such as the oak tables and Wegner chairs by Carl Hansen and the glass pendant fittings by Classicon, resonate with the essential and contemporary qualities of the architecture.

橡木桌子、卡尔·汉森（Carl Hansen）打造的椅子和 Classicon 的吊灯与建筑的现代风格相互呼应。

Groninger Museum Restaurant "Mendini"

格罗宁根博物馆"门迪尼"餐厅

Completion date: 2010
Location: Groninger Museum, Groningen, The Netherlands
Designer: Studio Maarten Baas
Photographer: Marten de Leeuw

完成时间：2010年
地点：荷兰 格罗宁根 格罗宁根博物馆内
设计：马腾·巴斯工作室
摄影：马腾·德·利乌

The Groninger Museum (originally designed by Alessandro Mendini, Michele de Lucchi, Coop Himmelb(l)au and Philippe Starck) underwent a major renovation in 2010. Jaime Hayon, Studio Job and Maarten Baas were commissioned to redesign, respectively, the interior of the information centre, the lounge and the restaurant.

For the very diverse and large group of guests of the restaurant 'Mendini', Maarten Baas developed an interior that functions as a café during daytime and that changes into an Italian 'a la carte' restaurant or is used for private parties in the evening.

2010年，格罗宁根博物馆（最初由亚历山德罗·门迪尼、米歇尔·德·卢基、蓝天组、菲利普·斯塔克共同设计）经历了一次较大的翻修工程。亚米·海因、Job工作室和马腾·巴斯分别受托负责信息中心、休息区和餐厅的设计。

考虑到餐厅不同类型的顾客，马腾·巴斯打造了一个独特的室内空间——白天可用作咖啡厅，夜晚则转换成一个可用于举办私人集会的意大利餐厅。

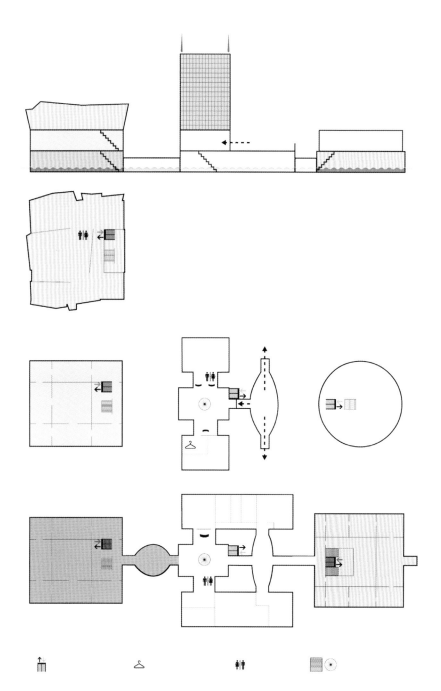

The interior is largely based on Maarten Baas' concept Clay furniture; the chairs and lights have all been clay-modelled by hand, making each object unique. Moreover, the red lights and the green chairs each have very subtle differences in colour, because of variations in the used pigments. The use of these pigments was especially developed for this interior, to make the furniture more user-friendly and better resistant to scratches. Also, the other objects, like the long benches and the bar, were made by hand, clearly displaying the recognisable signature of Maarten Baas. All furniture is of high quality and is created with great attention.

室内设计源自基于马腾·巴斯的粘土家具这一概念；椅子和灯都是由粘土手工制成，每件物品都别具特色。此外，由于使用的颜料的变化，每一个红色灯具和绿色椅子在颜色上都有细微的差别。为了使家具更易使用和耐刮擦，使用的所有颜料都是经过特别调配的。其他物品，例如长凳和吧台也是手工制作并带有清晰的马腾·巴斯的标志。所有家具都是高品质并精心打造的。

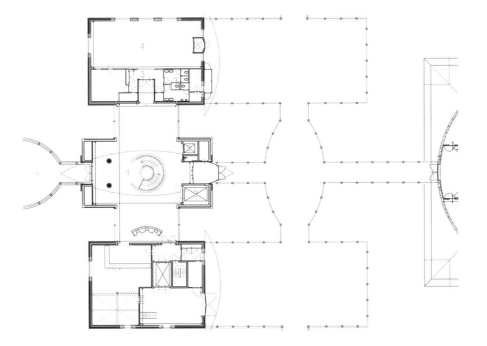

Despite the playful and unpredictable character of Clay furniture, the project as a whole is very spacious and open, because of the suppressed colour statement and the simple furnishing. Without losing its strong character, the restaurant forms a calm isle in the midst of all impressions that the Groninger Museum offers its guests.

粘土家具别具一格,并极具趣味性。质朴的色彩和简约的装饰使得空间更加宽敞。餐厅非但没有失去自身的特色,同时又在博物馆中营造出一个恬静的空间。

1. Museum hall   1. 博物馆大厅
2. Lift   2. 电梯
3. Corridor   3. 走廊
4. Dining room   4. 用餐区
5. Restaurant entrance   5. 餐厅入口

Floor plan   平面图

MOSI

曼彻斯特科学工业博物馆咖啡厅和餐厅

Completion date: January 2011
Location: Museum of Science and Industry, Manchester, UK
Designer: SHH
Photographer: Alastair Lever
Area: Café 195m², Restaurant 460m²

完成时间：2011年
地点：英国 曼彻斯特 曼彻斯特科学和工业博物馆内
设计：SHH建筑事务所
摄影：阿拉斯泰尔·利弗
面积：咖啡厅195平方米，餐厅460平方米

As part of the £9m refurbishment of Manchester's Museum of Science and Industry – which tells the story of Manchester's scientific and industrial past, present and future – SHH was commissioned to redesign the new hospitality offer, which includes a ground floor Caféteria and a first floor restaurant.

曼彻斯特科学和工业博物馆讲述了曼彻斯特科技和工业的过去、现在和未来，其翻新预算达900万英镑。SHH建筑事务所受托重新设计位于一层的咖啡厅和二层餐厅。

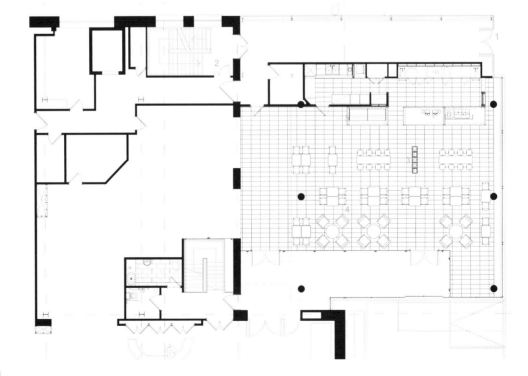

1. Entance
2. Staircase
3. Central counter
4. Dining and rest
5. Counter

1. 入口
2. 楼梯
3. 中央柜台
4. 就餐与休息区
5. 柜台

Ground floor plan　一层平面图

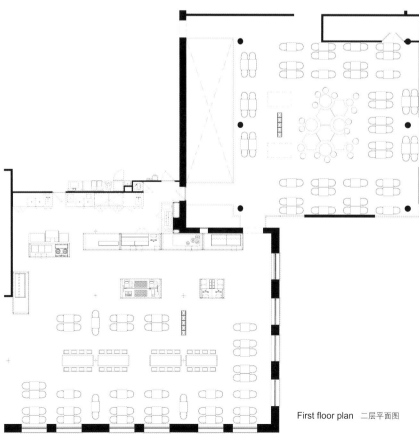

First floor plan 二层平面图

The design treatment was inspired by the Museum's industrial content and industrial feel and includes bespoke furniture and graphics.

设计的灵感来自于博物馆的工业展品和风格，包括定制家具和平面设计。

For the Café, a new graphic wall uses images from the MOSI archive to tell the story of historic food brands which began their life in Manchester.

A new 460sqm restaurant for Manchester's Museum of Science and Industry, as part of an overall £9m refurbishment project. The design treatment was inspired both by the Museum's content and by the old building's strongly industrial feel, including old brickwork, timber ceilings and huge steel structure.

Bespoke furniture included monolithic server tables – partway between industrial workshop tables and lab benches – with stainless steel tops and bright green steel 'section' legs.

A spillover area, aimed at school groups and families, includes bright and easily reconfigurable lightweight foam tables and angled, coloured stools.

在咖啡馆中，一个新的图片墙用博物馆档案中的图片讲述着那些从曼彻斯特起源，历史永久的食品品牌的故事。

这个460平方米的餐厅是曼彻斯特科学与工业博物馆900万英镑整体翻修工程的一部分。设计的灵感来源于博物馆的展品和老建筑的工业风格，包括旧的砖石结构，木质天花板和巨大的钢铁框架。

定制家具包括整体的桌子——在工业操作台和实验室长凳之间——有着不锈钢顶部和亮绿色的"分段"钢铁桌腿。

为了满足学校和家庭就餐团体的需要，设计师设计了明亮，易于重组的、轻质泡沫状桌子和角度、颜色多变的凳子。

Coach House, Hatfield House
门房改造餐厅

Completion date: 2012
Location: Hatfield House, Hatfield, UK
Designer: SHH
Photographer: Alastair Lever
Area: 557.4m²
Services provided: Interior design, branding & signage

完成时间：2012年
地点：英国 哈特菲尔德 哈特菲尔德庄园
设计：SHH建筑事务所
摄影：阿拉斯泰尔·利弗
面积：557.4平方米
设计内容：室内设计、品牌&标识

An existing tea-room for visitors to Hatfield House was already operating on the estate, with profits contributing to the ongoing costs of maintaining the historic property. Its simple format, however, was no longer equal to the demands being placed on it by the high volume of visitors. A new design was commissioned to enlarge the café and restaurant space and this coincided with a major master plan for the house and grounds, initiated by Lord Salisbury and undertaken by Brooks Murray Architects. The master plan was aimed at expanding alternative revenue sources, as well as improving year-round access to the grounds for visitors. The restaurant, housed in a former 19th century coach house, was to be joined by new retail facilities in the surrounding former stable buildings.

原有的茶室一直在经营，主要为方便庄园游客，而其盈利资金用于别墅维护。如今，其简单的风格和样式已不能满足日益增加的游客的需求。全新的计划包括咖啡厅和餐厅的扩建，而这一计划恰好与整体改建规划重合（由索尔兹伯里伯爵提出、布鲁克斯·梅尔建筑事务所负责执行）。整体改建规划旨在提高收入，同时满足游客需求。餐厅选址在建于19世纪内一座门房内，与邻近建筑改造的全新零售空间相连。

1. Glass extension and terrace
2. Ground floor dining area
3. Service counters
4. Restrooms

1. 玻璃扩展区与露台
2. 一层用餐区
3. 服务柜台
4. 卫生间

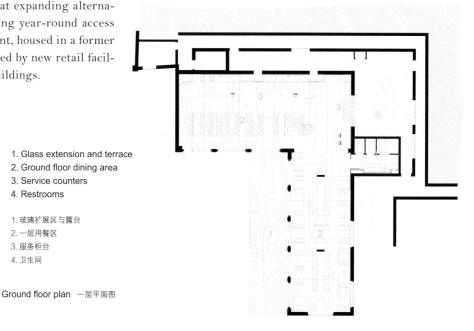

Ground floor plan　一层平面图

Ground floor seating area　一层就餐区

Ground floor space with stairs to first floor 一层就餐区内通往二层的楼梯

Cast aluminium spiral stairs 铸铝螺旋状楼梯

Back wall uses black-stained timber cladding 后墙采用黑色木材覆层饰面

The Deli with tiled counter 瓷砖饰面操作台

The Chef's Table 大厨操作台

Counter for The Bakery 烘焙柜台

Exposed red brick wall 裸露的红色砖墙

Bespoke communal tables with integrated display shelving 带有陈列架的定制长桌

Ground floor, extension and terrace　一层扩建部分和露台

Stairs up to first floor 通向二层的楼梯

Fontana Arte Chandeliers 丰塔纳艺术吊灯

First floor lighting detail 二层灯饰细节

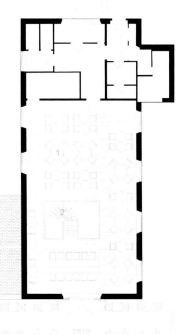

1. First floor dining area
2. Stairs

1. 二层用餐区
2. 楼梯

First floor plan 二层平面图

L'Osteria Künstlerhaus

艺术之家活动中心L'Osteria 意式餐厅

Completion date: 2011
Location: Munich, Germany
Designer: DiPPOLD Innenarchitektur GmbH
Photographer: DiPPOLD Innenarchitektur GmbH

完成时间：2011年
地点：德国 慕尼黑
设计：DiPPOLD装饰设计公司
摄影：DiPPOLD装饰设计公司

The historic Künstlerhaus (house of artists) is located near the old town of Munich. In former times used as a meeting place for artists and the upper class, it turned to a forum for Munich's citizens now.

After a complex restoration of the listed building, the Italian restaurant L'Osteria settled down in the front wing of the building. Without avoiding neither costs nor trouble, the first aim was to bring out the best of this place.

艺术之家坐落在慕尼黑老城附近，最初用作艺术家和上流社会名人的聚集地，如今改造成慕尼黑市民活动中心。

经过一系列繁复的改造工程，L'Osteria 意式餐厅落户在建筑前翼结构内。设计中避免考虑费用和任何其他困难，旨在充分利用这一空间。

1. Restaurant entrance
2. Bar
3. Kitchen
4. Venetian room
5. Dining room
6. Terrace

1. 餐厅入口
2. 吧台
3. 厨房
4. 威尼斯风格用餐区
5. 用餐区
6. 露台

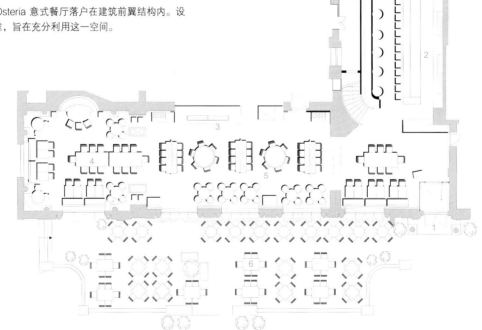

Floor plan 平面图

With about 180 seats, both in the restaurant and on the sun terrace, the L'Osteria at the Künstlerhaus offers daily Italian cuisine from 11.00 am till midnight.

餐厅设置189个座位（室内和露台），每天上午11点开始营业到午夜，主要提供意大利菜肴。

The red awning and the historical façade, dated back to the Wilhelminia time, welcome all pasta and pizza lovers.

A wall breakthrough to the kitchen, right in the centre of the main dining room, was a costly invest. By creating this breakthrough, guests are able to see 'behind the curtain' as they have an open sight towards the heart of the restaurant: the kitchen.

红色的遮阳篷和历史风格的外观让人不禁想起威廉明娜（前荷兰女王）时期特色，并欢迎着所有意大利面和比萨爱好者的到来。

主就餐厅中心区域内的一面墙壁被打破，花费大量的预算，旨在将餐厅的中心结构——厨房，呈现在顾客的眼前。

Another eye-catcher is the Venetian Room. The colossal chandelier combined with historical parts (ancient wall paintings) makes this room the most significant one within the L'Osteria.

Furthermore, all windows on the ground floor have been replaced with the effect lifting the border between the restaurant and the sun deck, in fine between inside and outside.

Comfort and flair was the designers' intention. By placing huge tables and dimmed light, there is nothing that can disturb a homelike time with friends and good food. Restaurant owner Friedemann Findeis and the interior architect Caroline Dippold agree: love in the detail and the passion for quality turn the L'Osteria into a unique restaurant.

餐厅内另一引人瞩目的空间即为威尼斯风格就餐区——硕大的吊灯和古老的壁画相结合，使其格外吸引眼球。

一层空间全部采用落地窗设计，拉近室内外空间的界限。

舒适和新颖是设计师追求的目标——宽大的桌子和柔和的灯光营造居家般的温馨氛围，使这里成为与好友聚会、享受美食的完美空间。餐厅老板弗里德曼·芬德斯（Friedemann Findeis）和室内设计师卡罗林·迪普德（Caroline Dippold）一致认为：对细节的热衷和对品质的热忱成就了L'Osteria餐厅的与众不同。

The Whitechapel Art Gallery Dining Rooms
白教堂画廊餐厅

Completion date: 2009
Location: The Whitechapel Art Gallery, London, UK
Designer: Project Orange
Photographer: Project Orange

完成时间：2009年
地点：英国 伦敦 白教堂画廊内
设计：Project Orange
摄影：Project Orange

The Whitechapel Art Gallery is the hugely influential London gallery that gave UK debuts to the likes of Picasso, Mark Rothko and Jackson Pollock. Project Orange was approached to design the new Dining Rooms as part of the gallery's ambitious expansion plans, doubling the size of the gallery by extending into the historic Victorian library building next door.

白教堂画廊在伦敦极具影响力，毕加索、马克·罗斯科和杰克逊·波洛克在英国的首秀都是在这里举办。画廊计划延伸到隔壁维多利亚图书馆处，而全新的餐厅则是扩建规划的一部分，由Project Orange负责设计实施。

The approach to the design of the Dining Rooms has been to try and forge a synthesis between the history and character of the original Arts and Crafts building, and the contemporary cutting edge of the exhibited works. In contrast to the expansive gallery spaces, these public areas are intimate and cosy, characterised by their timber panelling, pendant lighting and leather upholstery; with reclaimed library units contrasted with modern detailing, fixtures and fittings. The effect is to create a timeless space that seems at once modern and traditional, and where materials will change and improve with age and use.

设计尝试将古老艺术与工艺的历史和特色和现代时尚的艺术展品融合在一起。同开阔的画廊空间相比，餐厅更加亲切、温馨——木镶板、吊灯和皮质装饰更营造了舒适的氛围；古老的特色和现代的细节和装饰形成鲜明对比。最终打造了一个永恒的空间，传统与现代同在，材质随着时光的流逝不断改变着形象。

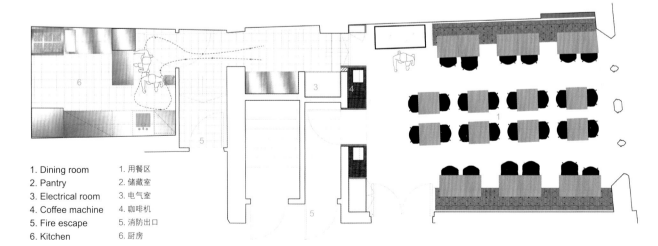

1. Dining room　　1. 用餐区
2. Pantry　　　　 2. 储藏室
3. Electrical room　3. 电气室
4. Coffee machine　4. 咖啡机
5. Fire escape　　 5. 消防出口
6. Kitchen　　　　6. 厨房

Main dining room plan　主用餐区平面图

1. Integrated banquette for less formal seating
2. Full height access to service riser
3. Lirica light fitting in polished brass
4. Fire exit
5. Service table
6. High level mirror
7. Wood panelling to match dining room
8. Informal banquette seating
9. Lucciole light in 'hell bronze'
10. Display shelving for books

1. 整体长椅用来弥补不足的正式坐席
2. 通往服务区的全高通道
3. Lirica 抛光黄铜灯具
4. 消防出口
5. 服务站
6. 高层镜子
7. 与用餐区相匹配的木镶板
8. 非正式的长椅坐席
9. Lucciole 灯具
10. 书籍展示架

Members dining room plan and elevations 会员用餐区平面图与立面图

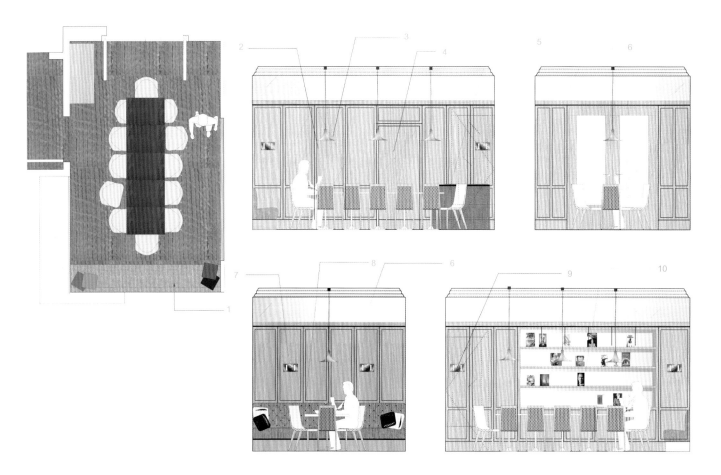

Members Dining Room "Bistro" 餐厅小包间

Chapter 3:
Dining in Theatres

第三章：剧院中的餐饮设计

3.1. General Design
总体设计

3.2. Design Requirements
设计要求

3.3. Functional Design
功能设计

Case Studies
案例赏析

Chapter 3: Dining in Theatres

3.1 General Design

There is to be an onsite service kitchen to receive catering prepared off-site that only requires final plating, presentation and serving. The ability to serve food to patrons and guests is required to Foyers, VIP Room, Rehearsal Rooms only with a secondary function to provide an efficient food service to the Green Room. Back of house access to the kitchen is required from the loading dock.

Beverage services are to be available in the Foyers and VIP Room. Scale of bar in each area suits capacity and service demand profiles. It is desirable that back of house support facilities are in close proximity, although not essential. Bars could be serviced from the front from an arrangement of central beverage stores in reasonable proximity of the loading dock.

This sector is comprised of the following facilities:
VIP lounge / boardroom
bar(s)
beverage coolroom
beverage store
service kitchen
coolroom
freezer
dry store
catering store
staffroom
uniform store
staff toilets & changerooms

3.2 Design Requirements

Back of House (BOH) service areas shall be grouped together with convenient access to loading dock for efficient movement of

3.1 总体设计

服务厨房将已准备好的食物装盘、陈列，并为大厅、贵宾室、排练室内的演员和顾客服务，同时为台下休息的演员提供食物。厨房后台应与运送食物装卸平台相同。

厨房为大厅和贵宾室供应饮品；酒吧区规格要求满足服务需求。辅助设施并不是硬性要求，但最好放在附近位置。酒吧区内的饮品供应集中配送，确保中央配送区与装卸平台相近。

这一区域包括如下设施：
贵宾休息室 / 会议室
酒吧区
饮品冷藏区
饮品中央配送区
服务厨房
冷冻区
干货储藏区
食物储藏区
员工室
制服存放区
员工卫生间 & 更衣室

3.2 设计要求

后台服务区要求与食物装卸平台入口相连，便于货品移动。贵宾休息室应远离剧院大厅，设有经由后台服务区而直接进入剧院的独立通道。

deliveries.
VIP Lounge shall be located off the main Theatre Foyer with direct access via BOH for discrete entry to venue by VIPs.
VIP room to have dedicated bar facility.
Bars servicing to public areas (foyers) should be grouped as far as practical to support an efficient human resourcing structure. Several bar facilities shall be provided if there is a concern that congestion problems and slow service will be a high risk if there is only a single bar.
Beverage service demand in the Studio Theatre Foyer is only expected to be a third of the Main Theatre Foyer – bar counter provisions shall be split accordingly.
Bar facilities are required to comply with requirements of the Department of Racing, Gaming and Liquor (DRGL).

3.3 Functional Design

3.3.1 VIP Lounge / Boardroom
① Requirements
Place for private entertaining of VIPs, including Government representatives, sponsors, patrons and the like.
Outside of VIP functions, space is to be available (flexible to serve as a room for meetings for up to 25 people).
Room to be conveniently accessible from main Theatre Foyer and BOH circulation system (discrete VIP access).

② Fitout
Occupancy: Up to 50 persons (generally standing) or 25 seated
Acoustics: Speech privacy to 45NR rating
Special Requirements: High quality environment
Floors: Resilient that is highly serviceable and easy to maintain
Walls: High quality finish
Ceilings: Min 2,700mm height
Doors/Access: Minimum 900mm width – BOH side
　　　　　　　Minimum 1,800mm width – BOH side
Outlook/Windows: Natural daylighting

贵宾休息室设有专门的酒吧区。
公共区域（大厅）内的酒吧区位置根据实际情况设定，满足顾客需求。应设置多个酒吧区，避免单独酒吧区带来的拥堵和延长服务时间的弊端。
剧场大厅酒吧区饮品供应量为剧院主厅酒吧区的三分之一。吧台长度按一定比例分割。
酒吧区设备要求符合酒精、博彩与赛马管理署相关规定。

3.3 功能设计

3.3.1 贵宾室 / 会议室
①要求
这是贵宾休闲区域，包括政府代表、赞助商、顾客等；除贵宾休息功能之外，还可行使其他功能（可容纳25人的会议室）；
确保空间从剧院大厅等空间可以直接进入。

②配备
空间容积：多达50人（站立）或25人（就座）。
音响效果：隔音效果需达到45NR（噪声比）等级。
特殊要求：高质量环境。
地面：耐用、易维护。
墙面：高质量饰面。
天花：高度不低于2.7米。
门/入口：后侧入口宽度不少于0.9米；
　　　　　后侧入口宽度不少于1.8米。
外观/窗户：自然光线照明。
　　　　　　良好视野。
陈列：最大化悬挂艺术作品。
设备：配备可伸缩数据投影仪和屏幕。
家具：供25人使用的桌椅。

Chapter 3: Dining in Theatres

Bar design in Deventer Schouwburg and Theatre Modernissimo
迪温特剧院酒吧与餐厅和Modernissimo剧院集市酒吧内吧台设计

High quality view
Display: Maximise opportunities for hanging artwork
Equipment: Retractable data projector & screen
Furniture: Lounge seating and coffee tables for 25

3.3.2 Bar(s)
① Requirements
To generally provide beverage services to patrons during intermission and for functions.
Ability to conceal bars (e.g. behind sliding screens).
Facility shall have the following bar counter lengths with equal back wall lengths:
Main Theatre Foyer 8m
Studio Theatre Foyer 4m
VIP Room 3m
Bar counter lengths exclude service and remain subject to detail design confirmation.
Stock storage and display requirements to future brief.
Beverage coolroom of minimum 2,400 internal width and with following doors:
Main Theatre Foyer Bar 10 x 2 tier
Studio Theatre Foyer Bar 4 x 2 tier
VIP Boardroom shall have a minimum 3.0m back wall beverage refrigeration unit with 4 x 2 tier doors.

② Fitout
Occupancy: Several staff per bar
Floors: Highly serviceable, slip-resistant and easy-to-clean/maintain resilient finish with integral coved skirting
Walls: Highly serviceable, washable and easy to maintain
Ceilings: Minimum 2,400mm height over work areas
Lower bulkheads acceptable over counters
Doors/Access: Minimum 900mm width
Security shutter over servery counter
Display: Opportunities to display products available and prices

3.3.2 酒吧区
①要求
中午休息期间为食客提供餐饮服务；
酒吧区可实现遮蔽（比如，可用滑动屏风遮蔽）。
不同空间内酒吧区柜台长度（后墙宽度）要求如下：
剧院主厅内：8米；
剧场大厅内：4米；
贵宾室内：3米。
吧台长度不包括服务空间，具体细节按要求而定。
存储区和陈列区设计要求：
饮品冷藏室内部宽度不小于2.4米，门的规格要求如下：
剧院主厅内酒吧区：上下两层，每层10个开门（10x2）。
剧场大厅内酒吧区：上下两层，每层4个开门（10x2）。
贵宾会议室内应安装不少于3米宽的冷藏设备，上下两层，每层4个开门（4x2）。

②配备
空间容积：每个酒吧区多名顾客。
地面：耐用、防滑、易清洗（采用一体凹圆踢脚线装饰结构饰面）。
墙面：耐用、易清洗、易维护。
天花：工作区高度不低于2.4米、低层隔板应在吧台之上。
门/入口：后侧入口宽度不少于0.9米；
　　　　服务柜台前安装安全帘。
陈列：陈列食品并显示价格。
设备：酒吧区应配备饮品冷藏柜、清洗玻璃杯的机器（每个酒吧区至少一台）、收银机。
家具：前台和后侧柜台、橱柜、不锈钢操作台、玻璃架。

第三章　剧院中的餐饮设计

Equipment: Fully equipped bars including:
Drinks refrigeration
Glass washers – 1 per bar
Cash registers
Furniture: Front and backwall counters, fronts and cabinets
Stainless steel work benches with necessary ice wells, spirit wells, glass racks, etc. as required to provide fully functional bar facilities.

3.3.3 Beverage Coolroom
① Requirements
Central main refrigerated storage for kegs and backup packaged stock at a temperature of 5°c.
To be located in BOH service area in reasonable proximity of bars and loading dock for efficient handling of stock deliveries and distribution.
② Fitout
Floors: Sealed, non-resilient
Walls: Insulated coolroom wall panels
Ceilings: Insulated coolroom ceiling panels
　　　　 Minimum 2,400mm height
Doors/Access: Minimum 900mm width. Door lock systems to code
Equipment: Refrigeration plant to mechanical consultant advice
　　　　 Beer system manifold by others
Furniture: Minimum 6m lineal length x 4 tier galvanized proprietary 4 tier shelving

3.3.4 Beverage Store
① Requirements
Central main storage for non-refrigerated beverage stock and consumables.
To be located in BOH service area adjacent to beverage coolroom and in reasonable proximity of bars and loading dock for efficient handling of stock deliveries and distribution.

② Fitout
Floors: Sealed, non-resilient

3.3.3 饮品冷藏室
①要求
中央冷藏区用于储存桶装饮品和包装饮品，温度保持在 5 摄氏度；
设置在后台服务区内，临近酒吧区和装卸区，便于饮品传递和分发。
②配备
地面：非弹性密封。
墙面：隔热冷藏室墙板。
天花：工作区高度不低于 2.4 米、隔热冷藏室天花板。
门 / 入口：后侧入口宽度不少于 0.9 米、门锁系统
　　　　　服务柜台前安装安全帘。
设备：机械顾问推荐的制冷设备、啤酒供应系统（多种类型）。
家具：直线长度不小于 6 米的 4x4 橱架。

3.3.4 饮品存储区
①要求
中央存储区，用于存放无需冷藏的饮品和食品；
设置在后台服务区内，临近饮品冷藏室、酒吧区和装卸区，便于物品传送和分发。

②配备
地面：非弹性密封。
墙面：耐冲击、易维护。
天花：工作区高度不低于 2.4 米、防穿入。
门 / 入口：宽度不少于 0.9 米。
设备：机械顾问推荐的制冷设备、啤酒供应系统（多种类型）。
家具：直线长度不小于 6 米的 4x4 橱架。

Chapter 3: Dining in Theatres

Walls: Impact-resistant and low maintenance
Ceilings: Minimum 2,400mm height, intruder-resistant
Doors/Access: Minimum 900mm width
Equipment: Refrigeration plant to mechanical consultant advice
　　　　　Beer system manifold by others
Furniture: Minimum 6m lineal length x 4 tier galvanised proprietary 4 tier shelving

3.3.5 Service Kitchen
① Requirements
Service kitchen for receiving, minor preparation, plating and distribution of food and dishwashing for catering associated with functions held within the venue. Primary preparation and cooking will be by a contractor located off site.
To be located in BOH service area in reasonable proximity of Foyers, VIP Room, Rehearsal Rooms and loading dock for efficient handling of deliveries and distribution.
Facility to comply with code requirements for commercial kitchens.

② Fitout
Occupancy: Approximately 5 kitchen staff
Acoustics: Acoustically isolated space to ensure internally generated noise is not transmitted to other occupied areas in the building
Floors: High serviceability, washable, easy to clean and maintain
Walls: High serviceability, washable, easy to clean and maintain
Ceilings: Minimum 2,700mm height, flush ceiling
Door/Access: Minimum 1,800 double door access
Outlook/Windows: Natural daylighting and external view
Display: Whiteboard

3.3.6 Coolroom
① Requirements
Short-term refrigerated food storage facility
To be located with direct access off Kitchen
② Fitout

3.3.5 服务厨房
①要求
服务厨房用于食物接收、准备、装盘和分发以及餐具清洗等。主要准备和烹饪过程应在其他场地完成。
设置在后台服务区内，临近大厅、贵宾室、排练室和装卸区，便于食物传递和分发；
满足商业厨房相关要求。

②配备
空间容积：约5名员工。
声响系统：封闭空间，避免厨房内的噪音传递到其他空间内。
地面：耐用、易清洗、易维护。
墙面：耐用、易清洗、易维护。
天花：工作区高度不低于2.7米。
门/入口：宽度不少于1.8米。
外观/窗户：双门入口，宽度不小于1.8米。
陈列：白板。

3.3.6 冷藏室
①要求
短期冷藏食物储藏设备；
设置在可直接通往厨房的区域内。

②地面：非弹性密封。
墙面：隔热冷藏室墙板。
天花：工作区高度不低于2.4米、隔热冷藏室天花板。
门/入口：入口宽度不少于0.9米、门锁系统。
家具：直线长度不小于6米的4x4橱架。

Floors: Sealed, non-resilient
Walls: Insulated coolroom wall panels
Ceilings: Insulated coolroom ceiling panels
　　　　　Minimum 2,400mm height
Doors/Access: Minimum 900mm width. Door lock systems to code
Furniture: Minimum 6m lineal length x 4 tier galvanised proprietary 4 tier shelving

3.3.7 Freezer
① Requirements
Short-term refrigerated storage facility for frozen foods
To be located with direct access via coolroom
② Fitout
Floors: Sealed, non-resilient
Walls: Insulated coolroom wall panels
Ceilings: Insulated coolroom ceiling panels
　　　　　Minimum 2,400mm height
Doors/Access: Minimum 900mm width. Door lock systems to code
Furniture: Minimum 4m lineal length x 4 tier galvanised proprietary 4 tier shelving

3.3.8 Dry Store
① Requirements
Central main non-refrigerated storage for kitchen stock and consumables
To be located with direct access off Kitchen
② Fitout
Floors: High serviceability, washable, easy to clean and maintain
Walls: High serviceability, washable, easy to clean and maintain
Ceilings: Minimum 2,400mm height, flush ceiling
Doors/Access: Minimum 900mm width
Furniture: Minimum 4m lineal length x 4 tier galvanised proprietary 4 tier shelving

3.3.9 Catering Store
① Requirements

3.3.7 冷冻室
①要求
短期冷冻食物储藏设备；
设置在可直接通往冷藏室的区域内。

②地面：非弹性密封。
墙面：隔热冷藏室墙板。
天花：工作区高度不低于2.4米、隔热冷藏室天花板。
门/入口：入口宽度不少于0.9米、门锁系统。
家具：直线长度不小于4米的4x4橱架。

3.3.8 干货储藏区
①要求
中央储藏区，用于存放无需冷藏的厨房使用材料和物品；
设置在邻近厨房的区域。
地面：耐用、易清洗、易维护。
墙面：耐用、易清洗、易维护。
天花：工作区高度不低于2.4米、隔热冷藏室天花板。
门/入口：入口宽度不少于0.9米、门锁系统。
家具：直线长度不小于4米的4x4橱架。

3.3.9 餐饮用具存储区
①要求
用于存放餐饮用具，如陶器、刀叉、玻璃器皿、设备和其他耗材等；
设置在邻近厨房的区域。
地面：耐用、易清洗、易维护。
墙面：耐用、易清洗、易维护。
天花：工作区高度不低于2.4米、隔热冷藏室天花板。

Chapter 3: Dining in Theatres

Storage facility for catering-associated crockery, cutlery, glassware, equipment and consumables
To be located with direct access off Kitchen
② Fitout
Floors: High serviceability, washable, easy to clean and maintain
Walls: High serviceability, washable, easy to clean and maintain
Ceilings: Minimum 2,400mm height, flush ceiling
Doors/Access: Minimum 900mm width
Furniture: Minimum 6m lineal length x 4 tier galvanised proprietary 4 tier shelving

3.3.10 Staffroom
① Requirements
Staffroom for catering services and usher staff
To be located in close proximity of BOH Food & Beverage Facilities
② Fitout
Occupancy: Up to 20 persons
Acoustics: Speech privacy
Floors: High serviceability, washable, easy to clean and maintain
Walls: High serviceability, impact-resistant, easy to clean and maintain
Ceilings: Minimum 2,700mm height
Doors/Access: Minimum 900mm width
Outlook/Windows: Natural daylighting and ventilation and external view
Display: Pinup and whiteboards to future brief
Equipment: Upright fridge
Microwave
Boiling water unit
Dispensing machines to future brief (at least 2)
Furniture: Dining tables and chairs for 25
3m kitchen bench with equal length of cupboard over for crockery storage

3.3.11 Uniform Store
① Requirements
Storage for catering and usher staff uniforms

门／入口：入口宽度不少于0.9米、门锁系统。
家具：直线长度不小于6米的4x4橱架。

3.3.10 员工区
①要求
用于餐饮服务和员工休息；
临近食物和饮品设备后台服务区。

②配备
空间容积：20人。
声响系统：完全隔音。
地面：耐用、易清洗、易维护。
墙面：耐用、防冲击、易清洗、易维护。
天花：工作区高度不低于2.7米。
门／入口：宽度不少于0.9米。
外观／窗户：自然光线、通风、良好视野。
设备：直立式冰箱、微波炉、烧水设备、自动贩卖机（至少2台）。
家具：用餐桌椅（25人）。
3米长的厨房操作台和同等长度并放置在餐具存贮柜上方的橱柜。

3.3.11 制服存放区
①要求
用于餐饮服务和员工制服存放。

②配备
地面：耐用、易清洗、易维护。
墙面：防冲击、易维护。
天花：工作区高度不低于2.4米。
门／入口：宽度不少于0.9米。
外观／窗户：自然光线、通风、良好视野。

② Fitout
Floors: High serviceability, easy to clean and maintain
Walls: Impact-resistant and low maintenance
Ceilings: Minimum 2,400mm height, flush ceiling
Doors/Access: Minimum 900mm width
Furniture: Minimum 4m lineal length of hanging rails

3.3.12 Staff Toilets and Changerooms
① Requirements
Toilets and changerooms for operational and catering services staff.
② Fitout
Occupancy: To code requirements for 40 staff
Acoustics: Acoustically isolate facilities – noise transfer to adjoining facilities not greater than 15NR
Floors/Walls: High serviceability, washable, easy to clean and maintain
Ceilings: Minimum 2,400mm height
Doors/Access: Minimum 900mm width
Equipment: Electric hand dryers
Furniture: Female toilets to have vanities (or equal)

设备：直立式冰箱、微波炉、烧水设备、自动贩卖机（至少2台）。
家具：用餐桌椅（25人）。
　　　直线长度不少于4米的悬挂结构。

3.3.12 员工卫生间和更衣室
①要求
为员工使用。

②配备
空间容积：40人。
声响系统：隔音，传递到其他房间的噪音不超过15NR（噪声比）。
地面/墙面：耐用、易清洗、易维护。
天花：工作区高度不低于2.4米。
门/入口：宽度不少于0.9米。
设备：电动干手器。
家具：女士卫生间应设有梳妆台。

Case Studies
案例赏析

Public Theatre Library Lounge
公共剧院休闲酒吧

BELG AUBE Tokyo Metropolitan Theatre
东京艺术剧场 BELG-AUBE 小酒馆

The Swan at the Globe Theatre
环球剧院天鹅餐厅

Barbican Lounge
巴比肯艺术中心休闲餐厅

The Bar & Restaurant, Deventer Schouwburg
迪温特剧院酒吧与餐厅

The Grand Café & Brasserie Pollux, de Maaspoort Theatre
Maaspoort 剧院咖啡餐馆

The Brasserie & Café, Theatre de Leest
Leest 剧院咖啡馆

Phantom Restaurant
幻影餐厅

Canteen Covent Garden
科芬园餐厅

Café Bar Theatro
剧院休闲酒吧

Bar Agora, Theatre Modernissimo
集市酒吧，Modernissimo 剧院

Cinepolis Luxury Cinemas - La Costa
Cinepolis 影院餐厅

Cinepolis Luxury Cinemas - Del Mar
Cinepolis 剧院餐厅

Paard van Troje
特洛伊木马大厅

Public Theatre Library Lounge

公共剧院休闲酒吧

Completion date: October 2012
Location: Public Theatre, New York, USA
Designer: Rockwell Group
Photographer: Emily Andrews, Noah Fecks
Area: 185.8m²
Seating: Bar and Lounge 84 seats 12 standing
Client: The Public Theatre
Services provided: architectural design, interior design, lighting design, custom furniture and fixture design

完成时间：2012年
地点：美国 纽约 公共剧院内
设计：罗克韦尔设计集团
摄影：艾米莉·安德鲁斯、诺亚·菲克斯
面积：185.8平方米
座区：84个座位 12个站位
客户：公共剧院
设计内容：建筑设计、室内设计、灯光设计、家具定制、固定装置设计

As part of a larger renovation of the Public Theatre, which includes a new façade, shopfront, entrance, lobby, mezzanine balconies and staircase, Rockwell Group has created a mezzanine bar and lounge. The mezzanine space did not exist before Rockwell Group's renovation, but thanks to a 25-foot open ceiling in the lobby, it was possible to create a cosy, almost hidden space within the void that was inspired by the idea of a vintage library.

罗克韦尔设计集团负责公共剧院翻新项目，包括外观、入口、大厅、中层露台和楼梯翻修。此外，设计师还在中层打造了一个新的酒吧和餐厅。翻修之前并不存在中层空间，设计师充分利用大厅内高达7.6米（25英尺）的天花，继而新建了一个温馨又几乎完全隐蔽的空间，灵感源自古老的图书馆风格。

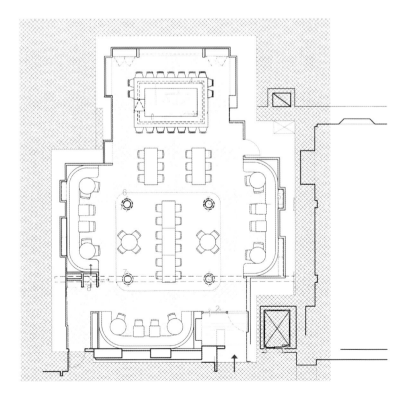

1. Recessed floor light
2. Decorative screen
3. Existing column
4. Bar stone counter top
5. Metal frame bookcase
6. Recessed floor light along column
7. New column to match existing column
8. Access door to mechanical unit & electrical panel

1. 嵌入式落地灯
2. 装饰屏风
3. 原有的柱子
4. 石质表面的吧台
5. 金属框架书柜
6. 沿着柱子排列的嵌入式落地灯
7. 与原有柱子相匹配的新柱子
8. 通向机械设备和电子面板的检修门

Floor plan 平面图

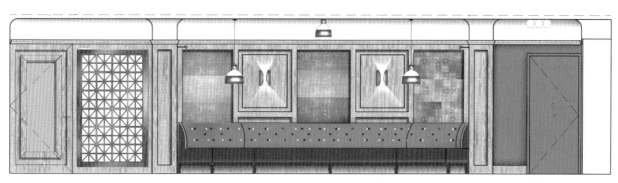

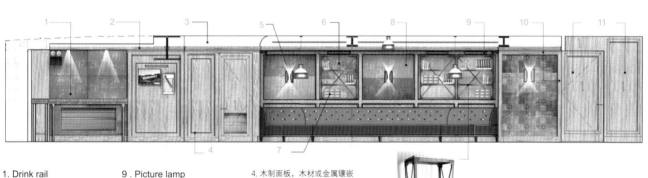

1. Painted gyp. cove
2. Wall sconce
3. Card catalog
4. Wall tile behind bookcase
5. Line of existing raw ceiling
6. Wood ceiling molding
7. Existing beam
8. Duct work
9. Duct work cavity

1. 彩绘石膏拱
2. 壁灯
3. 卡片目录
4. 书柜后面的墙砖
5. 原有天花板线
6. 木吊顶造型
7. 原有横梁
8. 管道工程
9. 管道工程孔穴

North elevation 北立面

1. Drink rail
2. Wood crown molding
3. GFRG cove
4. Wood panel w/metal inlay
5. Wall sconce
6. Wall tile behind bookcase
7. Card catalog
8. Wall tile
9. Picture lamp
10. 4" Wood trim & chair rail
11. Entry vestibule decorative wood panel & pocket door

1. 饮料围栏
2. 木制皇冠造型
3. GFRG 拱
4. 木制面板，木材或金属镶嵌
5. 壁灯
6. 书柜后面的墙砖
7. 卡片目录
8. 墙砖
9. 壁画灯
10. 4" 木饰和椅子围栏
11. 入口玄关，木纹装饰面板和折叠门

South elevation 南立面

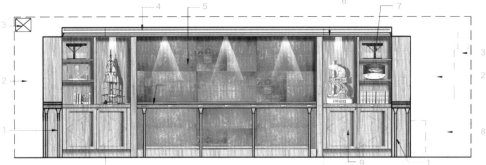

1. Metal leg
2. Duct work cavity
3. Mechnical duct work
4. Linear diffuser
5. Wall tile
6. Wood molding
7. Model display shelf
8. Return air duct
9. Lower cabinet

1. 金属腿
2. 管道工程孔穴
3. 机械管道工程
4. 线性扩散
5. 墙砖
6. 木制造型
7. 样本陈列架
8. 回风管
9. 低处橱柜

East elevation – back bar　东立面——吧台后部

Surrounding the gathering place are three niches defined by three button tufted banquettes set in front of industrial, steel-framed bookcases. The bookcases can be used to display vintage scripts, Shakespearean plays, or Theatre props. Wood-panelled walls also offer opportunities to display thematic content befitting the lounge, such as old posters from past Public Theatre performances.

餐厅四周是三个由纽扣图钉装饰的长座椅就餐区，摆放在工业风格的钢框书柜前。书柜用于陈列古代手稿、莎士比亚的剧本和剧院装饰物件。木板墙壁上也可用于摆放凸显餐厅主题的物品，如剧院之前演出的宣传海报等。

1. Wall tile
2. Picture lamp
3. Line of existing raw ceiling
4. Existing beam beyond
5. Table lamp
6. Wood molding cap
7. Bronze foot rail
8. Wood ceiling molding
9. Wood panel
10. Recessed EM light
11. Custom banquette seat

1. 墙砖
2. 镜画灯
3. 原有天花板线
4. 原有上方横梁
5. 台灯
6. 木制造型帽
7. 铜脚轨
8. 木吊顶造型
9. 木制面板
10. 嵌入式 EM 灯
11. 自定义长椅座位

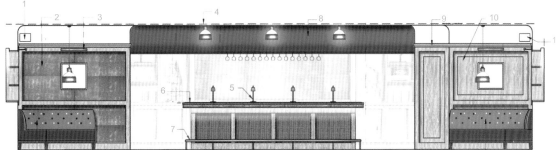

East elevation – bar　东立面——吧台

BELG AUBE Tokyo Metropolitan Theatre

东京艺术剧场BELG-AUBE小酒馆

Completion date: September 2012
Location: Tokyo Metropolitan Theatre, Tokyo, Japan
Designer: Aiji Inoue, Yuki Kanai / Doyle Collection Co., ltd.
Photographer: Satoru Umetsu/ Nacasa&Partners
Area: 54.07m^2
Client: M's Kitchen Inc.

完成时间：2012年
地点：日本 东京 东京艺术剧场内
设计：Aiji Inoue、优希金井/DOYLE COLLECTION 设计公司
摄影：梅津 聪/纳卡萨摄影公司
面积：54.07平方米
客户：M's厨房

BELG AUBE is a brasserie that mainly serves Belgium beer. It is placed inside the facility of Tokyo Art Theatre which is the origin of Art and Culture. Every day, the Brasserie is full of people who come to see Theatres and concerts and people who sponsor them.

In the design, the designers combined the individuality of Belgium beer and the comfort being suitable for the general so it will fit into the location in the best way.

First of all, variety is one of the characteristics of Belgium beer. To bring in the variety of beer into the design, they structured an open counter. The counter represents the characters of the brasserie directly. While on the other hand, stocks of beer are placed in the hanging cupboards. Here is the creative point. In order to stir the anticipation of customers, they chose to hide part of the beer and not show all the stocks. When viewed from different angles, one can get unequal glimpses of the arch-shaped window space. This will add another amusement aspect to the creative point as mentioned above. This is the biggest visual identity of the brasserie. Table seats are arranged in the public space. In total, the brasserie will look as if there are terrace seats. The space was intentionally created with a light atmosphere.

As for the colour scheme, by minimising the contrast, they have created usability. On the other hand, details are made with greatest care and used aging process to create constant high grade atmosphere.

BELG-AUBE 小酒馆位于东京艺术剧场内（艺术和文化的发源地），主要供应比利时啤酒。小酒馆内通常聚满来观看戏剧和音乐会的游客以及剧场的赞助者。

在设计上，设计师将比利时啤酒的特性和适于大众需求的共性结合，使其完美地融入到当地环境中。

首先，多样性是比利时啤酒众多特色之一。设计师将"多样性"运用到设计中，打造了一个开放式柜台，凸显小酒馆的特色。这也是整个设计的创意之一。为激起顾客的好奇心，设计师将部分啤酒遮挡起来——从不同的角度看去，可以瞥见拱形橱窗内的不同景象。这便构成了小酒馆的视觉特色。就餐桌椅摆放在公共区内，营造出露台般的感觉。

色彩选择以实用性为原则，尽量减少对比。细节设计被赋予更多的关注，打造出高品质的空间。

Terrace seats outside the facility give advertising effect to the people who walk by. 室外座区同时起到宣传作用，吸引着过往的行人驻足。

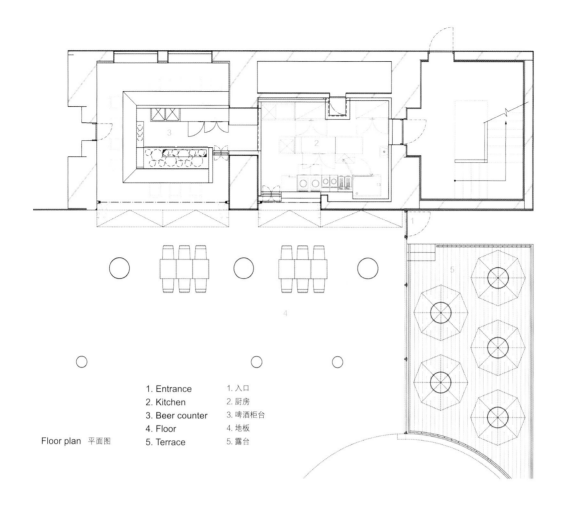

Floor plan 平面图

1. Entrance — 1. 入口
2. Kitchen — 2. 厨房
3. Beer counter — 3. 啤酒柜台
4. Floor — 4. 地板
5. Terrace — 5. 露台

The brasserie is separated into two parts: the public space and the private space. Though the public space is located indoors, it has an atmosphere of terrace seats. The bright colours with slight contrast make a comfortable place to enter.

Awning tent and accordion window bring open impression, and depth to the whole room.

Counter seats create an independent atmosphere.

Sophisticated counter designs match the impressions of Belgium beer's history. Variety of beer can be seen from the arch-shaped windows. This will capture the interest of the customers.

小酒馆被分割成两部分：公共区域和私密空间。公共区域设置在室内空间，但却营造出露台般的氛围。明亮而变化的色彩打造舒适的空间，吸引着顾客进入。

遮阳帐篷和手风琴状的窗户营造开阔的氛围，同时增添了空间纵深感。

柜台式座区营造独立的空间氛围。

精美的柜台与比利时啤酒历史相互呼应；品种繁多的啤酒在拱形橱窗内若隐若现，不断激起顾客的好奇心。

Functional looking counter and spacious seats make comfortable combination. Through the round-shaped window at the back of the room, one can see people walking by the main street located in front of the brasserie.

Ranges of different glasses are part of the design. The hanging cupboard has complex cross-section shape and combines both traditional and modern impressions.

功能性十足的柜台和开阔的座区完美结合。透过空间后部的圆形窗户，顾客可以看见大街上的行人。

不同形状的酒杯成为了设计的一部分；吊柜呈现出复杂的剖面结构，是传统与现代的融合。

Arched-shaped window space of the hanging cupboard 吊柜上的拱形橱窗

The Swan at the Globe Theatre

环球剧院天鹅餐厅

Completion date: 2007
Location: The Globe Theatre, London, UK
Designer: Brinkworth
Photographer: Brinkworth

完成时间：2007年
地点：英国 伦敦 环球剧院内
设计：Brinkworth室内设计公司
摄影：Brinkworth室内设计公司

Interior architect team Brinkworth has completed an exciting renovation of the restaurant and associated areas within Shakespeare's famous Globe Theatre on the South Bank of the Thames. With the partnership behind The Swan restaurant in Kent taking over the restaurants in the venue, Brinkworth have worked closely with Diccon Wright and Peter Cornwell to create contemporary, social spaces that retain a sense of place.

Kevin Brennan of Brinkworth says: "Given the context of a period-style building, an overly modern design would have felt incongruous. We wanted to create a concept that respected the history of the building in a contemporary way. I think that this is exemplified in the first floor restaurant. By marrying classic mid-century pieces with traditional materials the restaurant is given a sense of context, meaning it felt established from the minute it opened for trade."

Working within the parameters set by the Globe, Brinkworth has transformed the bar and restaurant spaces and has also created a flexible function room. Uniting the areas is a controlled use of materials, chosen to highlight the integrity of the natural state; oak, rusted metal and worn leather are staged against a warm, subtle colour palette of dark neutrals and white.

Diccon Wright of The Swan says: "Having worked in or been around restaurants for the last fifteen years or so, you get to know the style, the layout and format and who has designed what – looks get predictable. For Peter and myself, to choose Brinkworth ahead of two other 'more established restaurant designers' was a risk. Their approach was new and innovative and the results we believe different and exciting. Many finishes and styles established from their retail designs have successfully crossed over to the restaurant environment. The team are lovely people, creative and truly have a passion for what they deliver, which is again, quite novel within this industry!"

Brinkworth室内设计团队完成了位于泰晤士河南岸环球剧院内的"天鹅餐厅"的翻修项目，同迪肯·赖特（Diccon Wright）和彼得·康威尔（Peter Cornwell）共同打造了一个风格现代的社交空间，同时带有一定的地方特色。
来自Brinkworth室内设计团队的凯文·布伦南（Kevin Brennan）说："在一个古老建筑背景中，过于现代的设计往往不太和谐。我们力图构思一个理念，即用现代风格诠释古老历史。这在餐厅的设计中得到了完美体现——古典的中世纪物件与传统材质相结合。餐厅至此被赋予了一种独特的背景环境。"

设计师充分尊重环球剧院的背景环境，将餐厅和酒吧区进行改造，同时打造了一个灵活的多功能区。不同的区域之间通过材质而统一，凸显自然状态下的和谐特性。橡木、锈迹斑斑的金属、磨损的皮革与温馨的色调形成鲜明对比。

来自"天鹅餐厅"的迪肯·赖特谈到这一项目时说："我曾在餐厅中工作了15年，对餐厅的风格、布局和样式以及设计者一目了然。对于彼得和我来说，选择Brinkworth设计团队而并非在餐厅设计领域知名度较高的设计师无疑是一次冒险的行为。他们的设计方式新奇而富有创意，因此我们坚信最终的结果会是令人振奋的。他们将零售空间设计中运用的装饰和风格使用到餐厅环境中。不得不说，这个团队的成员非常可爱，他们富于创新精神，对设计极具热情。他们的设计方式在这一行业中也是别具一格的！"

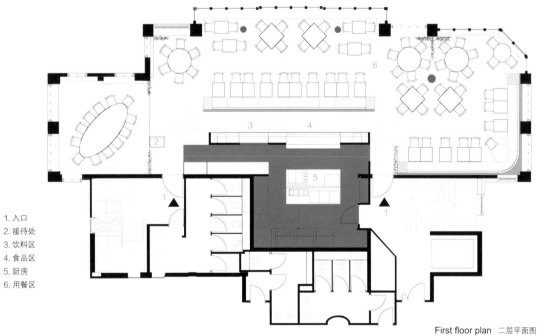

1. Entrance　　　　　1. 入口
2. Meet and greet stand　2. 接待处
3. Drinks pass　　　　3. 饮料区
4. Food pass　　　　　4. 食品区
5. Kitchen　　　　　　5. 厨房
6. Dining area　　　　6. 用餐区

First floor plan　二层平面图

The most formal of the spaces, the restaurant, is situated on the first floor, as entered from the riverside. Using a sophisticated colour palette of warm, mink grey and natural wood, the original flooring has been sanded back and stained a deep charcoal. Reflecting the theatre's history, old-fashioned ideas have been paired with contemporary materials in a Brinkworth designed, oversized armoire, wittily built from glass, wood and zinc. This has been cleverly conceived to provide ample room for two wait-stations, a bar and to highlight the chefs at work.

餐厅中风格较为正式的空间位于二层，从河边入口进入。空间主要采用温馨的珠光灰色和天然木材装饰，原有的地面经打磨之后饰以深碳色。设计师专门采用玻璃、木材和锌打造了一个超大橱柜，借以展现剧院的古老历史，并实现了古典风格与现代材质的融合。

The restaurant has been divided up by the imaginative use of Anaglypta wallpaper applied to ceiling panels. Painted mink grey and reminiscent of antique tiles, the three panels hover over the room creating intimacy without dictating the use of the space.

部分天花板采用浮雕壁纸装饰，在视觉上餐厅空间分隔开来。三块珠光灰色天花板悬浮在上空，让人不禁联想到仿古砖，并营造出温馨的氛围。

The possibility of a private dining area, seating twelve to fourteen, has been provided by a flexible, bespoke curtain, double sided to show soot black velvet on the public side and dark grey linen with an abstracted, white feather motif on the private side. The feathers reoccur throughout the project, inspired by The Swan's avian connotations and previously utilised by the restaurateurs in the Kent venue. The feathers continue around the perimeter of the entire room in the form of Ingo Maurer's 'Birdie' flying light bulb lamps.

White Eames Sideshell dowel – based chairs are used throughout the restaurant with variation coming from the mixture of arm and dining chairs. The smooth plasticity of the chairs is contrasted with the rusticity of the Benchmark designed, oak, pedestal tables used throughout. Reinforcing the concept of the use of natural materials, Brinkworth has created long, burnt oak sofas with worn, natural leather upholstery. Rusted metal legs enhance the level marble tabletops and Foscarini pendants hang above.

私密就餐区通过帘子与其他空间分隔开来，可容纳 12~14 人就餐。帘子朝向公共空间的一侧是烟灰色天鹅绒材质，另一则是带有白色抽象羽毛图案的深灰色亚麻布。羽毛图案主题灵感源自餐厅名称"天鹅"，贯穿整个空间。此外，羽毛化身成"小鸟"造型的灯饰，"飞翔"在餐厅内。

Eames "Sideshell"系列白色座椅被大量使用，包括扶手椅和餐椅。座椅光滑的表面与粗糙的橡木底座桌子形成鲜明对比。为深化"大量使用天然材质"这一理念，设计师运用橡木打造了长条形的沙发，并采用天然皮革装饰。锈迹斑驳的金属桌腿与大理石桌面和 Foscarini 吊灯形成对比。

On the ground floor is the bar, with a decadent, multifunctional marble bar allowing for food preparation and drinks service as well as hosting a crustacean tank that invites adventurous diners to view their dinner.

酒吧位于一层，颓废风格的多功能大理石吧台用于准备餐食，同时上面摆放着昆虫标本收藏罐，供敢于冒险的食客观赏。

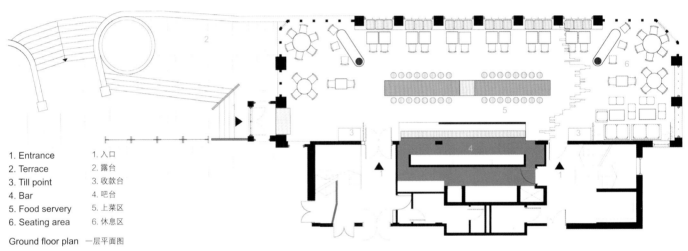

1. Entrance
2. Terrace
3. Till point
4. Bar
5. Food servery
6. Seating area

1. 入口
2. 露台
3. 收款台
4. 吧台
5. 上菜区
6. 休息区

Ground floor plan 一层平面图

Echoing the materials used in the restaurant, multiple mild-steel-based tables are identically designed apart from differing surface tops of burnt oak, natural oak and zinc. Inspired by The Swan team's concept of large, sharing tables, reminiscent of a familial, kitchen environment, Brinkworth has situated three large tables down the centre of the room. These are over-hung by an idiosyncratic lighting feature consisting of fifty bare, round light bulbs of varying dimensions hung at different heights from an Anaglypta ceiling panel, interspersed with The Swan's signature, Ingo Maurer designed, winged light bulbs. Benchmark has been commissioned to design, simple, concave, wooden stools that sit either side of the tables.

以软钢底座为基础的餐桌遍布整个餐厅，突出统一感；桌面风格多样，包括熏橡木材质、天然橡木和锌皮，营造个性化。餐厅经营团队对大型共享餐桌格外钟爱，营造出家庭聚会般的氛围。受到这一理念影响，设计团队专门选用了3张大型餐桌，由50个灯泡打造而成的独特灯饰悬垂其上，高低不同，而由Ingo Maurer设计的"天鹅餐厅"标识性灯饰穿插其间，营造出变化的空间纵深感。专门打造的简约的木质坐凳摆放在餐桌两侧，别具特色。

Enhancing the relaxed atmosphere of a domestic environment, vintage furniture has been used to create an informal lounge area. Mid-century, Danish, teak armchairs with careworn upholstery are juxtaposed with wickerwork dining chairs bought from the hip, guerrilla Reindeer restaurant that popped up for a month in Shoreditch during Christmas 2006, and reclaimed, oak, chapel chairs. The area is further defined by the use of different flooring materials. In the more relaxed lounge area long, narrow, stone tiles are used and gradually bleed into floorboards of similar dimensions in the dining area.

The perimeter of the room hosts vintage, stainless steel wall lamps and structural pillars commandeered to support high tables have been sanded back to the original nude-coloured concrete before being lacquered. This industrial effect is continued in the zinc, marble-topped wait stations and is softened by controlled use of feather-patterned curtains.

复古不锈钢壁灯装饰在餐厅四周墙壁上，用于支撑高脚桌的柱子结构被打磨成混凝土本色并经过喷漆处理，营造出工业风格。这一风格在锌皮和大理石饰面的等候区内进一步延续，羽毛图案窗帘的运用在一定程度了增添了柔和的气息。

为营造舒适的居家氛围，设计师选用复古家具装饰休闲区域，包括中世纪的丹麦风格柚木扶手椅、从 Reindeer 餐厅购买的柳条座椅和回收的橡木座椅。同时，地面采用不同材质装饰，窄长形状的石砖逐渐延伸到就餐区的地板处，更增添了这一区域的休闲风格。

Barbican Lounge
巴比肯艺术中心休闲餐厅

Completion date: 2010
Location: Barbican Centre, London, UK
Designer: SHH
Photographer: Gareth Gardner, Caroline Collett
Area: 250m²
Services provided: Interior design, furniture design, branding & graphics

完成时间：2010年
地点：英国 伦敦 芭比肯艺术中心
设计：SHH建筑事务所
摄影：加勒斯·加德纳、卡罗林·科利特
面积：250平方米
设计内容：室内设计、家具设计、品牌推广&平面设计

The 150-cover Barbican Lounge on the first floor, directly above the Foodhall offers small plate menus, as well as gourmet bar snacks and a special Dine & Dash menu, where diners can be out of the door in 50 minutes – perfect for those heading for a performance. The Lounge also features London's first Macaroon Mixologist, where the legendary French delicacies are twinned with a range of new and traditional cocktails.

The first floor Lounge has material links to its ground floor sister space, but also boasts a very individual and bold design treatment in striking colours, including peacock blue/green banquette seating with red upholstered buttons (using materials from Bute); vintage 1960s tables with murano glass tops and a variety of freestanding furniture in blue with splashes of green and red, including a Hans Wegner sofa. The touches of red, which also include the lacquered back panel of the new Stefan Bench/Helen Hughes chairs and a number of vinyl applications to glass panels, all reference Chinese lacquer red, historically used as a reference colour throughout the Barbican Centre.

巴比肯休闲餐厅位于二层（食堂之上），供应小盘菜、零食和特色速食菜肴，确保食客可以在50分钟之内就餐离开，适用于急于观看表演的顾客。餐厅内拥有伦敦第一个马卡龙调酒师，美味的法式甜点配以现代或传统的鸡尾酒，带来无与伦比的味觉享受。

休闲餐厅设计在材质选择上注重于一层空间呼应，但同时在色彩运用上彰显个性和大胆风格——孔雀绿的座椅点缀着红色的装饰扣子、20世纪60年代的蓝色玻璃面桌子和样式各异的独立家具饰以丝丝点点的红色和绿色。红色（中式丹红）同样被大量运用，并被赋予一定的参考意义。

1. Entry point　　　　　　1. 入口
2. Lounge seating　　　　 2. 休息区
3. Existing accessible WC　3. 无障碍卫生间
4. Bar　　　　　　　　　　4. 吧台
5. Theatre kitchen and food pass　5. 厨房和传菜区
6. Restaurant seating　　　6. 就餐区
7. Outdoor bar　　　　　　7. 室外酒吧
8. Outdoor lounge　　　　 8. 室外休闲区
9. Male WC　　　　　　　　9. 男士卫生间
10. Female WC　　　　　　10. 女士卫生间

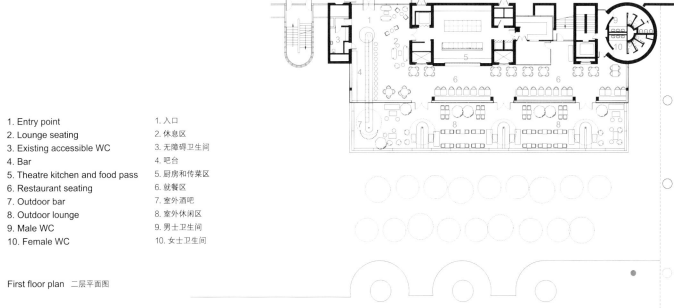

First floor plan　二层平面图

Perhaps the most stunning feature of the space, however, is the poured resin floor, also in peacock-green, created as an exact colour match of a Summer photo taken of the outdoor Barbican lake.

The Barbican's original hammered aggregate walls have been exposed in this space, which opens up to an outdoor terrace on one side and features two striking indoor walls to the right hand side on entry (the kitchen) and straight ahead (the bar). The kitchen is clad, floor-to-ceiling, in solid timber, with an open hot pass, whilst the striking 14m bar, which continues through the glazing onto the outside terrace, is in black glass with a black mosaic bar front.

餐厅内最令人惊奇的当属孔雀绿色树脂地面，恰似巴比肯湖面在夏季的色彩。

原有的骨料墙壁被裸露在外。餐厅一侧朝向室外就餐区，另一侧的室内墙壁格外引人注目。厨房全部（从地面到天花）采用实木饰面，长达14米的黑色玻璃吧台（黑色马赛克正面）一直延伸到室外露台区。

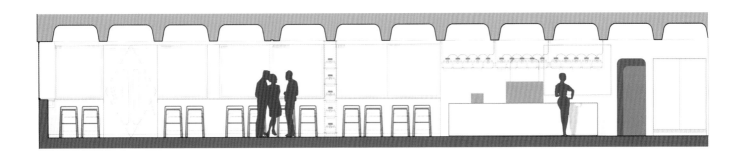

Close-ups of tables and chairs. 桌椅特写。

Lighting in the space is once again by .PSLAB. For the bar area, the lighting designers inserted black steel hoops, each carrying a clear halogen bulb topped with a brass circular reflector, so that the brass reflector serves in reflecting and directing the light. The longitudinal perception of the bar is highlighted by the repetition of the hoop-like insertions and their reflection in the glass façade of the terrace, acting as a mirror.

Over the dining area, .PSLAB developed a series of hoop chandeliers equipped with directional projectors within the fitting, making it functional for the dining space. .PSLAB felt it was important to keep an element of transparency through to the coffers, as the ceiling is such a prominent part of the space.

灯光设计由 .PSLAB 建筑事务所负责。在酒吧区内，设计师将卤素灯泡嵌入黑色钢圈内，并在顶端配置圆形黄铜反射结构，用于反射和指引光线。从水平方向看去，重复的圆环形状以及其在玻璃外观上反射的影子使吧台更加引人注目，犹如一面镜子。

在就餐区内，灯光设计师打造了一系列定向投影，增强其实用性。天花是空间中不可或缺的结构，而确保天花镶板的通透性至关重要。

The Lounge's outdoor terrace is linked to the terrace below by four of SHH's 'urban tree' umbrellas (with a further version used as a waiter station at the far end). Planting (again by Kate Gould) includes plants that trail over the terrace to mimic the residents' window boxes (which have always featured trailing plants), and seven large concrete pots featuring olive trees, with lavender at their base, to continue the exotic and strongly coloured dynamic design treatment for this space.

On the terrace, object fixation and mounting was only permitted onto the long planter running along the side of the terrace perpendicular to the bar. The position of the exterior lighting elements therefore along the planter reflected the axis of the mullion on the glass façade. The lighting elements consist of conical heads made from brown folded metal sheets, fixed to stainless steel rods. The detail connecting the head to the rod is a short neck articulation, allowing a multitude of lighting orientations.

Outdoor seating on the terrace includes white lounge wire chairs, white wire tables (bespoke-designed by Helen Hughes of SHH), timber benches and also Robin Day's black polo chairs.

"This is a light reference to Robin Day, the original art director of the Barbican back in 1972", explained Helen Hughes. "We wanted to pay our respects, without being in danger of taking the project down memory lane. Here and throughout, it was important to respect the Barbican's structure and history, but also to create a fresh and unified scheme, which makes its own truly contemporary statement for now and for the years to come."

室外就餐区与露台之间通过四个命名为"城市之树"的伞形装饰连通。露台绿化以蔓生植物为主，七个大型混凝土花盆内栽种着橄榄树，底端则种植着薰衣草，以室内设计风格相互呼应。

设备和其他装置要求只能安装在沿露台与酒吧区垂直方向的花架上；沿花架一侧的照明结构反射出玻璃外观上的窗棂。照明结构包括固定在不锈钢金属棒上的由褐色弯曲金属片打造的锥形帽。其中值得注意的细节是，金属棒和锥形帽之间通过可旋转结构连结，便于改变照明方向。

室外就餐区包括白色休闲铁丝椅、白色铁丝桌、木凳和黑色座椅（polo chair）。"这一设计是向巴比肯艺术中心艺术总监罗宾·戴致敬，海伦·修斯解释说："我们致力于向其致敬，但避免将设计置于回忆的漩涡。尊重巴比肯艺术中心原有的结构和历史至关重要，但我们也尽力营造一个清新并与之相和谐的空间，使其在当下和未来成为现代风格的诠释。"

The Bar & Restaurant, Deventer Schouwburg
迪温特剧院酒吧与餐厅

Completion date: 2011
Location: Deventer Schouwburg, Deventer, The Netherlands
Designer: M+R interior architecture
Photographer: Studio de Winter / Herman de Winter
Materials: Special red leather; special white natural acrylic stone Alpine White (LG HI-Macs); mirror HPL; smoked oak wooden floor

完成时间：2011年
地点：荷兰 迪温特 迪温特剧院内
设计：M+R室内建筑设计公司
摄影：赫尔曼·德·温特/温特工作室
材料：特质红色皮革，白色亚克力人造石，HPL镜子，烟熏橡木地板

The Deventer Schouwburg is the city's stage, the meeting place for producers and visitors, and as such it's important that everyone feels welcome. So M+R designed an open and multifunctional interior. The transparent façade makes the large lobby on the ground floor seem like a continuation of the public area. With one theatrical movement the two high revolving doors, which are made of glass, sweep visitors onto the stage. The bar, cloakroom and restaurant are positioned in such a way that the lobby can also be used for presentations or receptions and the like. Also along the stairs there are generous, organically shaped stages/sitting areas that form an integral part of this area. Large sliding doors allow the lobby to be used separately, enabling a more efficient use of the theatre. The style of the lobbies on the first and second floor has been modified and they have been given a new bar and a raised platform.

迪温特剧院是城市的舞台，是劳动者和访问者的聚集地，每个人在这里都可以感受到温馨。设计师打造了一个开放的多功能室内空间，通透的外观使得一层的大厅犹如公共空间的延续，两扇宽敞的玻璃旋转门恭候着参观者的到来。酒吧、衣帽间和餐馆位置精心规划，大堂可同时用作接待厅。沿楼梯两侧是宽敞而有序的休息区，成为公共空间的一部分。滑动门的设计使大厅可以单独使用，增添了剧院的灵活性。二层和三层的大厅风格加以改善，增添了新的酒吧和被抬升的平台。

True eye-catchers are the large round revolving doors through which visitors will enter the theatre.

When entering the hallway the first thing one sees are the stairways in which the Superior Covelight warm white is installed. The organic shaped elevated stages/seatings are lit by Covelight Superior RGB, which is installed in the steps.

Specially formed bars with HI-MACS, was the material chosen and it is present in all functional furniture of the public area: 1 mobile helpdesk, three service desks and an impressive curved white main bar.

两扇宽敞的玻璃旋转门欢迎着参观者的到来，格外引人注目。

走进大厅之后，首先映入眼帘的便是通向上层的楼梯，白色凹面灯装饰在两侧，别具特色。这一灯饰也用于装饰可升高的舞台。

人造石被广泛使用在特殊形状的吧台上、移动服务台、固定服务台和曲线造型的白色吧台上。

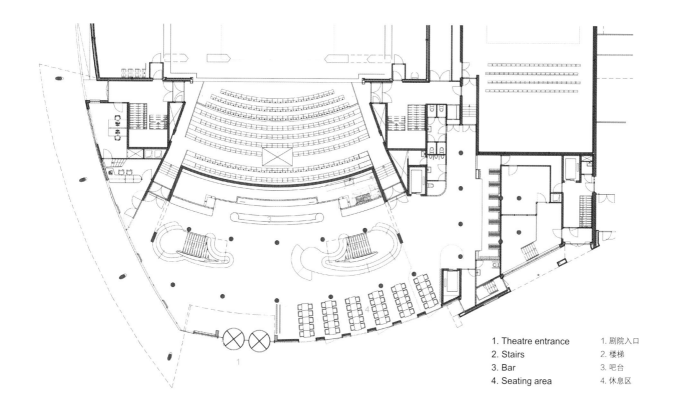

1. Theatre entrance 　1. 剧院入口
2. Stairs   　2. 楼梯
3. Bar   　3. 吧台
4. Seating area   　4. 休息区

Ground floor plan　一层平面图

The 9.5 metre-long white curved bar is a ship shape, designed as a tribute to Deventer's history as a river harbour. It comprises five different sections, which are seamlessly joined to create the iconic bar.

长达 9.5 米的白色曲线状吧台模仿船舶的造型，旨在向迪温特曾作为河港的历史致敬。吧台包括 5 个不同的部分，连结在一起，形成一个别具特色的结构。

Deventer is originally a trading city located at the river 'de IJssel': a historical Hanzenstad with a rich history in inland shipping. That's why M+R designed the bars in the theatre as ships. The appearance of the bar is a white hull: a mineral material that consists acrylic resin.

The bar is partly located under the large theatre hall.

To mask the sloping ceiling and to create a spatial effect a mirror ceiling is applied.

迪温特曾是位于艾瑟尔河（river 'de IJssel'）岸边的贸易城市，拥有丰富的内河船运历史。这也是设计师将吧台打造成轮船造型的原因。吧台外观犹如白色空壳，采用富含丙烯酸树脂的矿物材质打造。

酒吧的一部分位于剧院大厅下方。

为掩盖天花的坡度并营造出开阔感，设计师采用镜面装饰天花。

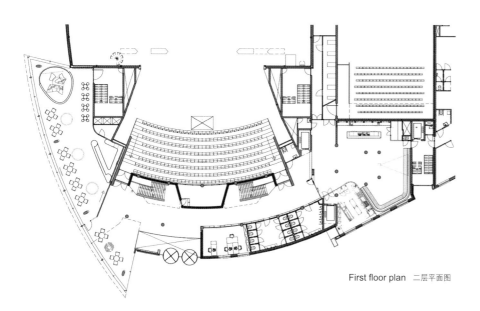

First floor plan 二层平面图

Although the theatre is flexible and the foyers can be used as banking areas, one foyer is divided and used as a special restaurant area. The kitchen is also located in the foyer area on the first floor. The interior design is contemporary: the wooden floor is extended in all areas so that no distinction is made between the different spaces. The big columns are also applied with mirrors. The walls are covered with metal chain curtains that are specially lit.

尽管剧院的空间比较灵活并且门厅可以被用来提供金融服务，一个门厅被分隔成了一个特别的餐厅区域。厨房也位于二层的门厅区域。室内设计充满现代感：木质地板在各个空间延伸，因此不同的空间没有差别。巨大的柱子也安装了镜子。墙壁上覆盖着特别光亮的、由金属链条组成的窗帘。

The Grand Café & Brasserie Pollux, de Maaspoort Theatre

Maaspoort剧院咖啡餐馆

Completion date: November 2011
Location: de Maaspoort Theatre, Venlo, The Netherlands
Designer: M+R interior architecture
Photographer: Studio de Winter

完成时间：2011年
地点：荷兰 芬洛 Maaspoort剧院内
设计：M+R室内建筑设计公司
摄影：温特工作室

The 'Grand Café' located on the ground floor is directly accessible from the marketplace for the theatre. The glass façade gives a beautiful interaction between inside and outside. The façade will be opened during spring and summer time, so the internal space becomes a part of the market.

咖啡厅位于一层，可从剧院前的市场直接进入。玻璃外观在室内外之间建立了一种关联，在春天和夏天完全敞开，因此咖啡厅成为了市场的一部分。

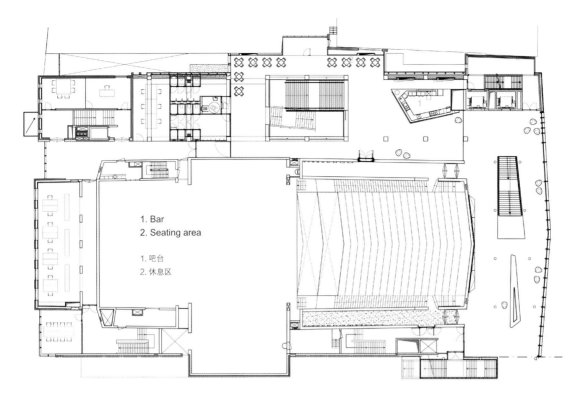

First floor plan 二层平面图

1. Bar
2. Seating area

1. 吧台
2. 休息区

The interior conceptual approach is the river that floats nearby the theatre. The interior is predominantly warm white with purple accents composed of natural sustainable materials. The design is inspired upon water and shipping elements, such as the bars.

餐厅室内设计方式源自剧院附近的河流，温馨的白色构成主色调，紫色天然环保材质点缀其中。水流和航海元素构成主要灵感，如吧台的设计。

In the foyers here and there are boulders brought by the river. The floor is like a wooden Jetty on the embankment surrounded by running water.

There is also a contemporary new brasserie named Pollux. The furniture and materials are white; guests give colour to the environment.

大厅内摆放着巨大的卵石,犹如从河岸冲过来的一般。地面犹如被水环绕的木码头。

同时还有一个叫做波勒克斯的新酒馆。家具和材料为白色,客人们为空间增添色彩。

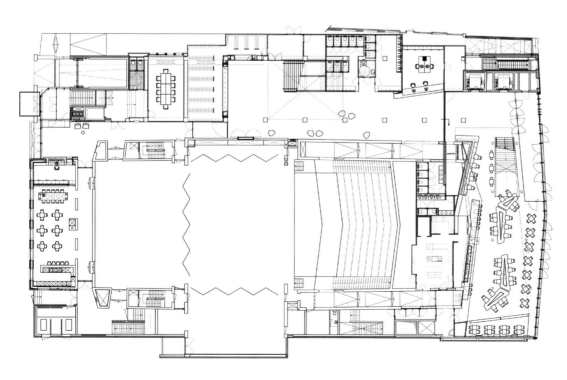

1. Bar
2. Mobile benches
3. Dining area

1. 吧台
2. 移动桌椅
3. 就餐区

Ground floor plan 一层平面图

The mobile benches are designed as small sailing boats which are docked. The benches divide the space, providing for the guests intimacy as a bay on the coast line. To create space the benches have seating place at both sides.

The floor is like a ship deck composed of Keroewing parquet.

Above the bar are suspended lighting fixtures 'Hope' by luceplan. The prism glass parts give a special lighting reflection upon the ceiling like the reflection of the sunlight on running water.

移动的座椅犹如停泊靠岸的小帆船,将空间分隔开来,同时营造出亲密的氛围。为打造独特的空间感,座椅两侧都设有座位。

木质拼花地板犹如小船甲板一般。

吧台上方悬垂着 luceplan 制造的灯饰"希望之光",棱柱玻璃结构将光反射到天花上,犹如阳光洒在水面上。

The Brasserie & Café, Theatre de Leest

Leest剧院咖啡馆

Completion date: September 2010
Location: Theatre de Leest, Waalwijk, The Netherlands
Designer: M+R interior architecture
Photographer: Studio de Winter

完成时间：2010年
地点：荷兰 瓦尔韦克 Leest剧院内
设计：M+R室内建筑设计公司
摄影：温特工作室

In 1996, the doors of the theatre opened for the first time to visitors. 14 years later, the theatre has become a complete makeover. The entrance has a large revolving door so the entrance space in the lobby could be added to the foyer. The box office and retail space and cloakroom has been converted into a brasserie with a large open bar. The theatre cafe is a combined function and is also used as ticket counter. As guests come to order their theatre tickets they can stay and enjoy also a range of food and drinks.

1996年，Leest剧院首次向游客开放。14年后，这里发生了天翻地覆的变化。入口增添了宽敞的旋转门，将门厅延展到中央大厅内。售票处、商店和衣帽间改造成了带有开放式吧台的小餐馆。剧院餐馆同时行驶着售票处的功能，顾客前来买票的时候，便可以在这里惬意地小食一番。

The entire foyer features a new resin floor and the 'fuchsia' colour is adjusted in black, white and gold; this colour scheme is also implemented in the theatre.

A new spiral staircase is connecting the ground floor with the above situated area which is designed as a lounge area.

整个门厅采用了全新的树脂地面,"紫红"的色调是由黑色、白色和金色调配而成;这个配色方案也应用在了剧院之中。

一个新的螺旋式楼梯连接着一层和上方被设计成休息区的空间。

1. Foyer
2. Bar
3. Dining area

1. 门厅
2. 吧台
3. 就餐区

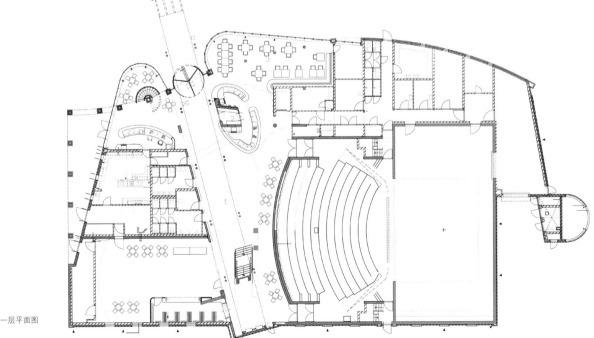

Ground floor plan 一层平面图

By creating an organic-shaped bar as a connecting element between the foyer and restaurant area, the bar leads the bar the visitor on a natural way to the brasserie.

Waalwijk is a well-known city in The Netherlands for her shoes production. The shoe shapes have served as a basic concept for the design of the bars appearance, which has a leather finish. Through indirect lighting the shapes are accentuated.

The brasserie is a flexible space with tables and chairs; only the big wall-mounted bench is fixed. The bench also shapes the space and the soft leather look provides softness.

Special light fixtures that are integrated in the wall covering are used as signing.

一个有机形状的酒吧被用来连接门厅和餐厅区域；使观众自然地进入酒馆。

瓦尔韦克在荷兰以制鞋业闻名。酒吧的外观设计以鞋的形状为基础，并以皮革装饰。通过使用间接照明，酒吧的形状得到了突出。

酒馆中的桌椅灵活摆放，只有安装在墙壁上的长凳是固定的。长凳也塑造了空间，柔软的皮革外观给空间带来了柔软感。

安装在墙面上的特制灯具被用作酒馆的标识。

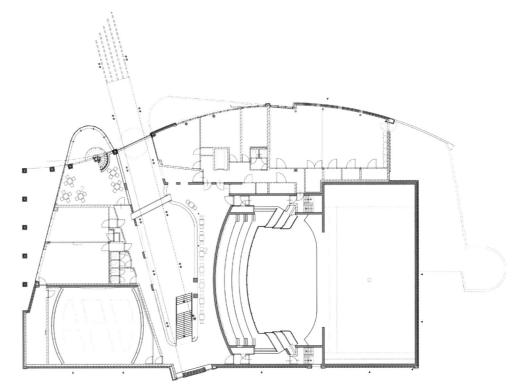

1. Spiral stairs to the first floor
2. Seating area

1. 通往二层的螺旋楼梯
2. 休息区

First floor plan 二层平面图

Phantom Restaurant

幻影餐厅

Completion date: 2011
Location: Opera Garnier, Paris, France
Designer: Studio Odile Decq
Photographer: Roland Halbe
Area: 1,300m²

完成时间：2011年
地点：法国 巴黎 加尼叶歌剧院内
设计：奥蒂尔·德克工作室
摄影：罗兰德·哈尔伯
面积：1,300平方米

Like a 'phantom', silent and insidious, the soft protean curves of the mezzanine level float above the dinner guests, covering the space with a surface that bends and undulates. Behind the columns of the east façade of the Opera Garnier, the restaurant is located in a place where, when the building first opened, horse-drawn carriages would drop off ticket holders, arriving for a performance. Creating a new space in the Opera Garnier meant following strict guidelines concerning the historical character of the monument: in order to ensure the possibility of completely removing the project without damage to the existing structure, the designers were not allowed to touch any of the walls, the pillars, or the ceiling.

柔和而又多姿多彩的曲线仿佛悬浮在食客上方，沉静而又神秘，犹如幻影一般。曲线结构一直向下延伸，构成空间表面。加尼叶歌剧院在刚刚营业的时候，游客通常坐着马车前来观看表演。餐厅选址在这里意味着设计中要严格遵循古典建筑准则。为确保避免对原有结构造成损害，设计师被要求必须保留原有墙面、梁柱和天花结构。

The façade of the restaurant is a veil of undulating glass, sliding between each pillar. With no visible structure, the glass is held in place by a single strip of bent steel running along the arched curve of the ceiling. This steel strip is fixed to the upper cornices of the columns 6 metres above the ground with stainless steel connecting rods. The glass is therefore held in place as if 'by magic'. The façade therefore allows for clear views and a minimum impact. Providing enough floor space to seat 90 people was another requirement for this limited space. The mezzanine was therefore created as a continuous surface. Narrow columns extend upwards towards the molded plaster hull, which curves to form the edges of the handrail. This vessel, which has been slipped under the cupola, is a cloud formation floating between the existing elements of the room without touching them. It's an allusion to the changing form of the phantom, whose white veil glides surreptitiously in space.

餐厅外立面由连续的玻璃打造，位于梁柱之间。玻璃通过沿天花曲线方向弯曲钢条结构固定，而钢条则通过不锈钢连杆固定在距离地面6米之上的檐口上。因此，从表面看来，玻璃似乎以一种不可思议的方式固定着。透过玻璃，可以清晰地看见室外的景象。设计中面临的另一挑战是在有限的空间内打造出可容纳90人就餐的座区。夹层犹如一个连续的表面，细长的梁柱向上延伸到石膏模铸结构处，构成栏杆的边缘。这一结构位于圆屋顶之下，犹如云朵一般悬浮在原有结构之间。

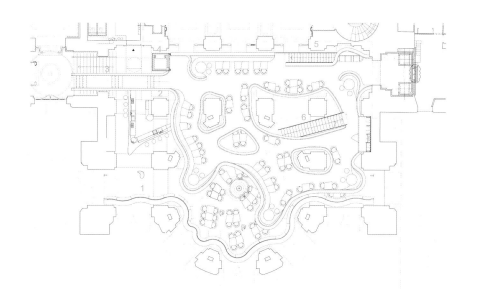

1. Glass façade
2. Bar
3. Ramp
4. Lift
5. Service stairs
6. Main stairs
7. Mezzanine

1. 玻璃外观
2. 吧台
3. 斜坡
4. 电梯
5. 服务楼梯
6. 主楼梯
7. 夹层

Mezzanine floor plan　夹层平面图

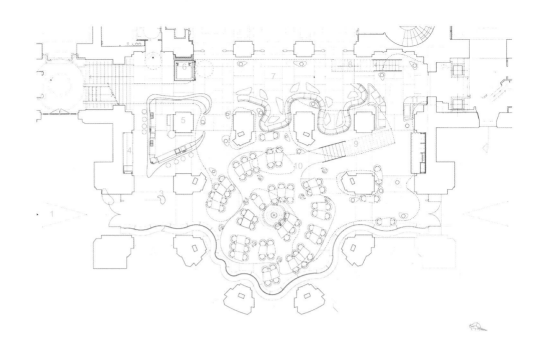

1. Restaurant entrance	1. 餐厅入口
2. Glass façade	2. 玻璃外观
3. Reception	3. 接待处
4. Cloakroom	4. 衣帽间
5. Bar	5. 吧台
6. Lift	6. 电梯
7. Lounge	7. 休息室
8. Service stairs	8. 服务楼梯
9. Main stairs	9. 主楼梯
10. Restaurant	10. 餐厅
11. Office	11. 办公室

Ground floor plan 一层平面图

Quietly, almost insidiously, the soft protean curves of the mezzanine cover the space with a volume that arches, undulates, and floats above the guests. The space is open and turned outward. The keystone of the existing dome remains visible from the ground floor, while suddenly becoming very close to the diners on the upper level. Sitting close to the stone arches of the ceiling, the symmetry of the cupola is no longer apparent, the reference points change, and sense perception of the space is altered.

In the curve of the hull above, immersed in warm red tones, the upper level becomes an intimate and private space.

夹层内布满了柔和而多变的曲线，蜿蜒起伏，悬垂于食客上方。空间格外开阔，建筑原有拱形结构依然清晰可见，在上层就餐的食客似乎可以触手可及。近距离接触天花的石头拱形结构之后发现，圆屋顶的对称性已悄然逝去，空间感也在悄悄地改变。

曲线结构内点缀着温暖的红色，上层空间俨然成为了一个亲切而私密的区域。

The red carpeting flows down the steps of the main staircase dramatically, spreading out into the centre of the black floor below, and running under the tables until it arrives at the edge of the façade.

At the back of the room, in the area closest to the entrance to the Opera, the space becomes more protected and private, contrasting with the whiteness of the rest of the room. Long red booths line this space, creating the 'lounge' area for the restaurant. At the outside edge of the lounge, a long black bar snakes around a nearby column. The design for this project is based around creating a space that will highlight the restaurant inside the Opera Garnier, without mimicking the existing monument, but respecting it while affirming its truly contemporary character.

红色的地毯顺着楼梯台阶一直向下延伸到楼下黑色地面中央，穿过桌椅到达空间的边缘。

靠近剧院入口处的空间更加私密，与其他以白色调为主的空间形成对比。红色的长座椅占据了这一空间，使其成为餐厅内的休闲区。与这一区域临近的空间内，长长的黑色吧台蜿蜒在柱子附近。这一设计的目的是凸显餐厅空间，摒弃但却尊重原有建筑元素，呈现代特色。

Canteen Covent Garden

科芬园餐厅

Completion date: June 2012
Location: Lyceum Theatre, London, UK
Designer: Very Good and Proper
Photographer: Very Good and Proper
Capacity: restaurant and bar 130 covers

完成时间：2012年
地点：英国 伦敦 Lyceum大剧院内
设计：Very Good and Proper
摄影：Very Good and Proper
座位：餐厅&酒吧共130个座位

Following the success of Canteen's temporary 'restaurant in residence' in Covent Garden's Piazza (over 800 people a day in a 70-cover site), the popular mini-chain confirms that another permanent restaurant – its fifth – is about to open its doors.

Canteen Covent Garden is located on the ground floor of the Lyceum Theatre where currently one of the most popular shows in London is on stage, The Lion King. The entrance on Wellington Street is to the right of the theatre and is one of the gateways to Covent Garden that attracts over 50 million visitors a year. The restaurant and bar design reflects Canteen's now recognisable signature style, contemporary, democratic and welcoming.

The site may be new but the ethos remains the same: all day dining from breakfast to dinner, with lunch and afternoon tea in between, enabling customers to enjoy reasonably priced food that is good quality, freshly prepared and seasonal.

Canteen owners Dominic Lake and Patrick Clayton-Malone opened their last Canteen in 2009, in Canary Wharf. They have waited to find the perfect location for their fifth site, and with its enviable location at the Lyceum Theatre and light and spacious bar and dining room, Canteen Covent Garden is a welcome addition to this burgeoning and exciting area of London.

"Our expansion strategy has been highly considered and measured. This is the right time and, most importantly, the right site to expand the Canteen restaurant group." – co-founder, Dominic Lake

"It's exciting to be part of a new wave of high-quality eateries that have recently opened in Covent Garden and to be able to serve our Great British Food to Londoners, theatre-goers and an international audience." – co-founder, Patrick Clayton-Malone

继科芬园临时餐厅（70个座位，每天接待800多名顾客）取得成功之后，这一小型连锁餐饮空间开设了另一家永久性餐厅（也是其第五家餐厅）。

科芬园餐厅位于Lyceum大剧院一层。威灵顿大街入口位于剧院右侧，也是通往科芬特公园的入口之一，每年游客数量多达5000万。就餐区和酒吧区的设计凸显现代风格，温馨气息十足。

选址是全新的，但餐厅一直秉承的理念却依然不变：供应早、中、晚餐和下午茶，让顾客花费合理的价钱享受高品质、新鲜、应季食物。

连锁餐厅所有者多米尼克·莱克（Dominic Lake）和帕特里克·克莱顿－马隆（Patrick Clayton-Malone）曾于2009年在金丝雀码头（Canary Wharf）开设第四家餐厅。之后一直为其第五家餐厅寻找合适的地址，最终选定Lyceum大剧院。科芬园餐厅成为伦敦这一蓬勃发展的地域内的备受欢迎的场所。

"我们的扩建计划经过仔细考虑和权衡，我们选择了恰当的时机和正确的地点。"——多米尼克·莱克

"作为高品质餐厅的一员，并为伦敦人、剧院游客和来自世界的观众供应英式美食是值得令人兴奋的事情。"——帕特里克·克莱顿－马隆

Canteen is celebrated not only for its all day menu but also for its progressive approach to design. The restaurant group designs and produces its own furniture, through British design studio Very Good & Proper. The 'Canteen Utility Chair' is hugely popular, and is found in each Canteen restaurant, its design and style now instantly recognisable. However, the chair is not solely for Canteen or just restaurant use; it can now also be found in sought-after locations around the world – MoMA Sweden, BBC and Channel 4 creative meeting spaces, and the new Facebook headquarters in San Francisco. The chair is sold through leading international design retailers. The Canteen Utility Chair has been redefined for Canteen Covent Garden with brushed brass metalwork.

With the original façade of the building dating from the 1830s, Canteen's contemporary design is juxtaposed with the ornate molding around the tall windows and front doors.

餐厅备受欢迎不仅仅是因为其全天候供应食物，更因其循序渐进的空间设计方式。餐厅自己设计并打造家具，其中知名的"Canteen实用座椅"（Canteen Utility Chair）广受欢迎，并应用到其连锁餐厅内，其设计风格大受认可。这一座椅目前不仅仅出现在餐厅内，在BBC和Channel 4的会议中心内也可见它的身影。目前，这款椅子经过改装并采用金属铜饰面。

原有建筑立面始建于19世纪30年代。扩建餐厅被赋予现代风格，与原有的古典装饰相互补充。

The bar area is sleek and inviting with its zinc bar top. Very Good & Proper designed bar stools and chairs and reclaimed Victorian mosaic floor tiles.

酒吧区时尚而温馨。锌皮表面的吧台、专门打造的吧凳和椅子及回收维多利亚风格马赛克地砖格外吸引眼球。

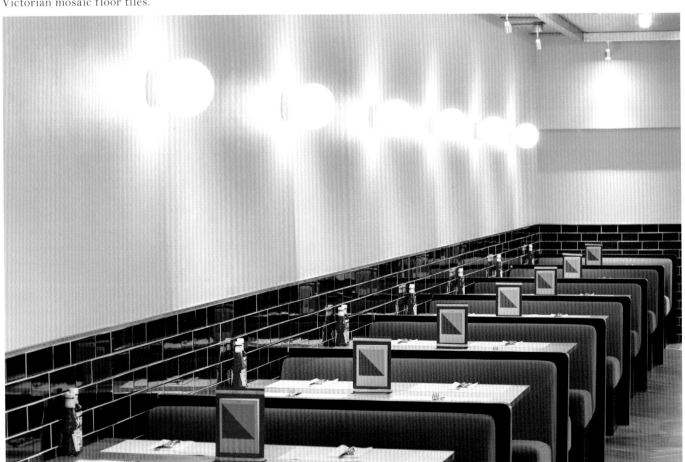

A slope leads down to the large oak herringbone-floored dining room with booth seating lined against a tiled wall (original tiles used by London Underground); the 'Covent Garden Chair' in various colours sits at the round tables, while the brushed brass Canteen Utility Chair lines the long tables beneath the windows.

一条坡道通往就餐区——橡木铺设人字形地板别具特色，皮卡座沿着瓷砖墙壁摆放，餐厅特有的座椅被饰以多种颜色并摆放在圆桌旁，黄铜刷面处理的座椅摆放在窗户下面。

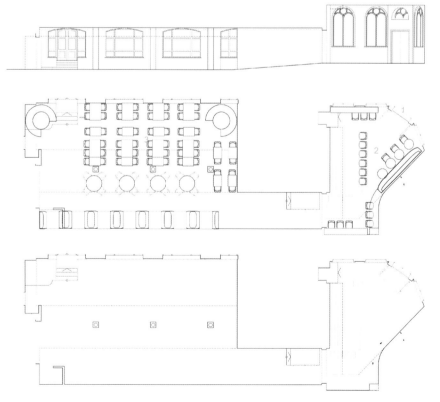

1. Restaurant entrance
2. Bar
3. Dining area

1. 餐厅入口
2. 吧台
3. 就餐区

Floor plan 平面图

Café Bar Theatro
剧院休闲酒吧

Completion date: 2009
Location: Komotini, Greece
Designer: Minas Kosmidis
Photographer: Studiovd N.Vavdinoudis–Ch.Dhmhtriou
Area: 280m²

完成时间：2009年
地点：希腊 科莫蒂尼
设计：米纳斯·科斯米迪斯
摄影：Studiovd N.Vavdinoudis–Ch.Dhmhtriou
面积：280平方米

This is a lounge bar with an intense taste of tailor-style elegance and delicate post-industrial influences. It reveals a 'theatrical stage' consisting of cosy straw couches, sculptural tables, metallic lighting fixtures, black and white digital prints, frames with plaid fabrics and mirrors on the side. The whole space is organised as a scene where clients both act and observe. All the above are designed and placed in harmony in order to provide a comfortable, almost homey atmosphere, through warm colours on the walls, furniture and natural materials which will make visitors feel free to relax and enjoy their drink within the hustle and bustle of the city.

这是一家休闲酒吧，呈现定制的优雅和精致的后现代风格。酒吧犹如剧院的舞台一般，摆放着舒适的草编沙发和餐桌。其中一侧，金属灯饰、黑白数码装饰画和格子图案的镜框别具特色。整个空间恰似安排有序的场景，客人即是看客又是演员。暖色调的墙壁和家具及天然材质打造了如家般的环境，让客人远离城市的喧嚣，在这里尽情的放松和畅饮。

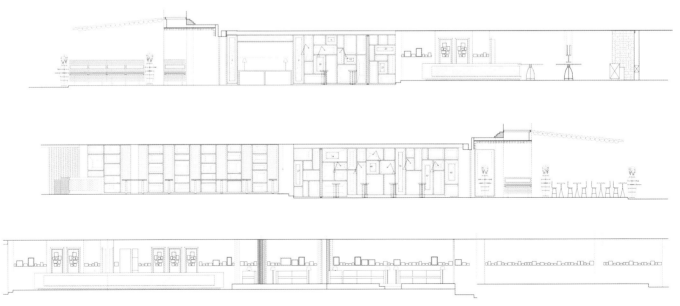

Sections 剖面图

1. Entrance　　1. 入口
2. Main bar　　2. 主吧台
3. Secondary bar　　3. 副吧台
4. Elevated lounge　　4. 抬高的休息区

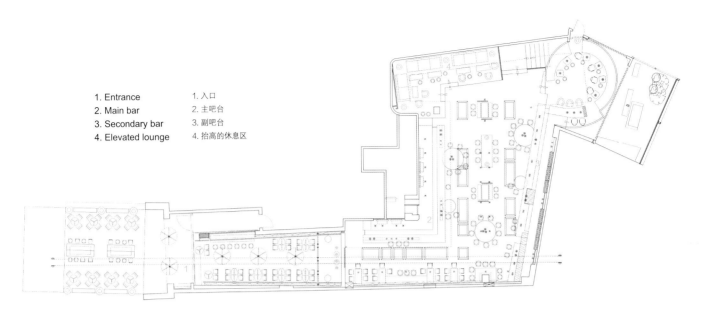

Ground floor plan　一层平面图

With memories of the atmosphere of Manhattan and Hollywood at the 1920's and 30's decades, combined with contemporary references and correlations, was erected the new scene.

新的装饰风格取材于20世纪20年代和30年代的曼哈顿与好莱坞，并加以相关的引述和回忆。

With yellow as the main colour, metallic luster and the shine of the black and white Stars of that era, the area acquires Class-Chick aesthetics.

黄色作为空间内的主色调，似乎回到了那个"星"光闪烁的年代，并被赋予特有的古典美。

古典风格的旧家具被再次利用。

The old furniture reinvested

Velvet and silk fabrics with dots are used as finishing touches to metallic bronze leather and metallic fabrics.

天鹅绒和丝绸织物摆放在青铜色皮革沙发上,起到装饰作用。

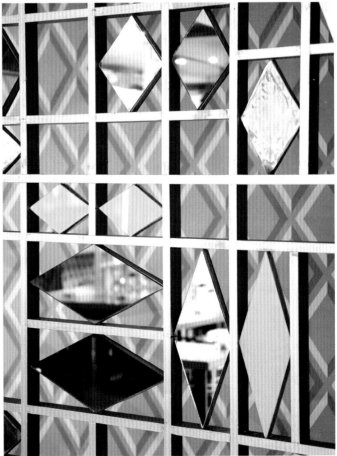

Wallpapers on all walls and voile curtains supplement the scenery. 墙壁上张贴的壁纸和纱帘是对这个空间场景恰到好处的补充。

Bar Agora, Theatre Modernissimo

集市酒吧，Modernissimo剧院

Completion date: March 2010
Location: Bergamo, Italy
Designer: Gritti Architetti
Photographer: Gritti Architetti

建成时间：2010年
地点：意大利 贝加尔莫
设计：Gritti建筑设计事务所
摄影：Gritti建筑设计事务所

In 2005 the municipality decided to begin a renovation of the civil centre, with the aim to re-establish the original relations between the two opposite buildings placed on the shorter sides of the Piazza della Libertà. Both of the buildings were projected by Alziro Bergonzo; originally designed to Opera Nazionale Balilla and to Casa del Fascio, they become during time the cinema-theatre Modernissimo and the city hall.

2005年，市政府决定改造文化娱乐活动中心，旨在在自由广场（Piazza della Libertà）两侧矗立的建筑之间构筑一种关联。两幢建筑最初分别为巴利拉组织（Opera Nazionale Balilla）和法西奥大楼（Casa del Fascio），最终改造成影剧院和市政厅。

The bar is located next to entrance hall and it has an independent entrance to assure the functioning of the drink even if the audience is closed.

酒吧紧邻入口大厅，拥有自己独立的入口，确保剧院关闭时也能正常开放。

The bar counter is totally covered by glass.

吧台完全采用玻璃材质打造。

Details of the bar counter. 吧台细节。

The walls are covered by bamboo wood with featured lighting. 墙壁采用竹子饰面并安装了特色照明灯。

Cinepolis Luxury Cinemas - La Costa
Cinepolis影院餐厅

Completion date: February 2012
Location: Carlsbad, California, USA
Designer: SMS Architects, Irvine, California
Photographer: Errol Higgins
Area: 2,777m²

完成时间：2012年
地点：美国 加州 卡尔斯巴德
设计：SMS建筑事务所
摄影：埃罗尔·希金斯
面积：2777平方米

Aiming to revolutionise the movie-going experience, Cinepolis Luxury Cinemas converted an existing aged movie theatre into a contemporary luxury brand 532-seat 6-screen cinema.

这一项目的主要目的旨在提升观影体验，因此将原有的古老影院改造成包括532个座位的奢华影院。

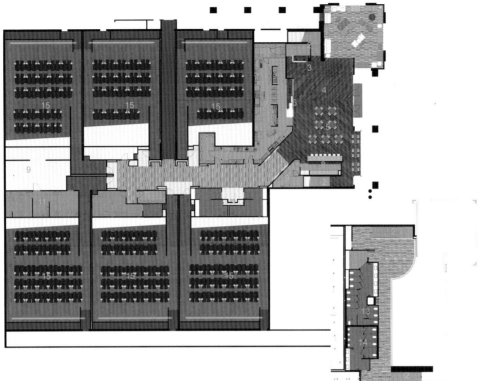

1. Lounge
2. Stairs
3. Concierge
4. Lobby
5. Dining
6. Concessions
7. Kitchen
8. Bar
9. Storage
10. Promenade
11. Offices
12. Employee lounge
13. Women's restroom
14. Men's restroom
15. Auditorium

1. 休息室
2. 楼梯
3. 门房
4. 前厅
5. 就餐区
6. 特许经营区
7. 厨房
8. 吧台
9. 储藏室
10. 走廊
11. 办公室
12. 员工休息室
13. 女卫生间
14. 男卫生间
15. 放映厅

Ground level & upper level 一层 & 二层平面图

The new design addresses the need to create a gathering space for a new cinema experience and expands the potential of the facility by providing food and drink services in an upscale restaurant atmosphere.

Ticket sales are encouraged to be bought on line through the cinema web site or at the lobby's touch screen monitors which allow movie previews reserved seating options. Wi-Fi is also available for patrons and a concierge desk provides assistance to theatre customers.

全新的设计着重强调打造一个能够体现观影体验的集会空间，拓展影院的潜力，提供餐饮服务。

顾客可以通过影院网站或大堂内的触摸屏幕购票，直接选择座位。这里提供无线网路服务，同时服务台为顾客提供其他服务。

The open lobby dining and bar, along with a quiet comfortable lounge invites customers in and creates a relaxed atmosphere. By utilising wood floors, contemporary furnishing and lighting, a zone of casual elegance that encourages the customer to relax and enjoy the space has been created.

Keeping service and comfort in mind, this innovative theatre renews the cinema experience with personal touches – an elevated level of customer service and gourmet movie fare form a blend of hospitality and cinema to develop a sense of a special night out for the customer.

The modernised interior continues throughout the building including theatre auditoriums that feature high-quality state-of-the-art digital equipment, HD, 3D and surround sound technologies, plush seating with dining trays and concierge call buttons. The updated cinema along with its first class service creates an undisturbed atmosphere for enjoying movies, special events, good friends and great food.

餐吧位于开敞的大堂内,安静而舒适的氛围吸引着顾客的到来,为其提供轻松的就餐体验。木质地板、现代风格的装饰和灯饰更增添了些许的随意与典雅。

良好的服务和舒适性是一直秉承的宗旨,影院通过独特的个性诠释观影体验——高档的顾客服务和合理的票价为顾客带来美好的观影之夜。

影院放映厅内以现代装饰风格为主,高品质的设备一应俱全。此外座椅上配置餐牌和订餐服务按钮,方便顾客点餐。高档的氛围和一流的服务为顾客营造一个观影、会友和享受美食的完美场所。

Cinepolis Luxury Cinemas - Del Mar

Cinepolis剧院餐厅

Completion date: July 2011
Location: Del Mar, California, United States
Designer: SMS Architects, Irvine, California
Photographer: Errol Higgins
Area: 3,251 m²

完成时间：2011年
地点：美国 加州 德尔马
设计：SMS建筑事务所
摄影：埃罗尔·希金斯
面积：3251平方米

As the centre piece of the extensive re-visioning of the Del Mar Highlands Centre, this 8-screen theatre remodel offers the ultimate in the movie palace experience and marks the entry of Cinepolis Luxury Cinemas into the United States.

剧院是德尔马中心地区的焦点，提供奢华的影院体验，同时成功地将Cinepolis引进美国。

At approximately 35,000 square feet the theatre will offer 8 stadium auditoriums, all digital, with 559 fully reclining leather lounge chairs.

剧院面积约为3251平方米,包括8个阶梯式观众礼堂,共559个皮质活动靠背座椅。

Guests are treated to an upgraded lobby and expanded food and beverage service including alcoholic beverages. Waiters serve the guests in the seating areas of the lobby or within their reserved seats within the auditoriums. The finishes offer the feeling of a high-end hospitality experience.

这里提供餐饮服务,并可供应酒精饮料。顾客可在大堂休息区内或礼堂内就餐,高档的装饰营造出高端的就餐体验。

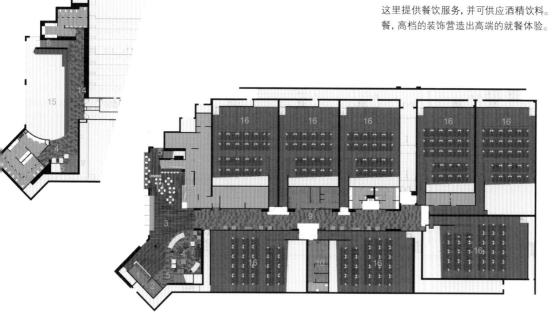

1. Box office	1. 售票处
2. Storage	2. 储藏室
3. Lobby	3. 前厅
4. Lounge	4. 休息室
5. Stairs up to balcony	5. 通往楼厅的楼梯
6. Concessions	6. 特许经营区
7. Kitchen	7. 厨房
8. Bar	8. 吧台
9. Promenade	9. 走廊
10. Employee rooms	10. 员工房间
11. Men WC	11. 男卫生间
12. Women WC	12. 女卫生间
13. Dining	13. 就餐区
14. Balcony	14. 楼厅
15. Lobby below	15. 向下通往前厅
16. Auditorium	16. 放映厅

Level one & balcony   一层 & 楼厅平面图

213

Paard van Troje
特洛伊木马大厅

Completion date: 2012
Location: The Hague, The Netherlands
Designer: studio BARBARA VOS and studio TILIA
Contractor/Interior contractor: De Timmerij
Structural engineer: Konstructieburo Snetselaar BV
Building physics consultant: AaCee Bouwen en Milieu
Installation advisor: Steegman and TIBN
Photographer: Stijnstijl Photographer
Area: 450m^2
Client: Het Paard van Troje, International Concert Hall

完成时间：2012年
地点：荷兰 海牙
设计：芭芭拉·沃思工作室与椴树工作室
承包商/室内设计承包商：德·第默里奇
结构工程师：斯奈特斯拉尔工程公司
建筑工程顾问：AaCee环境建造公司
安装指导：斯蒂格曼与TIBN
摄影：斯蒂金斯蒂尔摄影
面积：450平方米
客户：特洛伊木马国际音乐厅

After forty years, in the summer of 2012 the famous concert hall, 'Het Paard van Troje' (the Trojan Horse) was thoroughly refurbished. The duo from The Hague, studio BARBARA VOS and studio TILIA were appointed to transform the foyer of the concert hall into the public living room of the city. The design, typified as 'raw glistening' adheres to the changing needs of concert going public; a night out does not only needs to be a grand event but also an intimate and personal one.

Ever since the last refurbishment ten years ago 'Het Paard van Troje' has become a significant concert hall in The Netherlands and abroad. Its maze of corridors, stairs and halls were redesigned by OMA Rem Koolhaas. The last ten years visitors' needs have changed, adding to the clients' wish to create a comfortable bar and seating area in which people can have a quiet drink or a small snack next to the two existing stages. Celebrating its forty-year history 'Het Paard van Troje' has been significantly refurbished in the summer of 2012.

The present day also calls for alternative ways of financing projects in the cultural sector. In the spring of 2012 'Het Paard van Troje' was offered for sale in a special way. Part of the costs for the refurbishment has been covered by crowd funding. Through buying a share the public could not only participate in the refurbishment, but also permanently become part of that refurbishment in the form of their name prominently shown on glass. The famous band from The Hague, DI-RECT, has acquired the complete smoking room in this way.

The design's biggest asset is its use of industrial materials combined with its simple appearance. Strong elements of the original design have been cleverly accentuated to better experience its robust spatial character. The concrete construction has been painted in a light colour, non-load-bearing walls were clad in untreated birch wood with the floor being poured in a rough brown composite.

The intimate living room character of the bar and smoking room, contrasts with the robust atmosphere of the rest of the foyer through the use of smoked oak floorboards. Large yellow-painted springs stand out while carrying the small stage. Cables have all been directed over the rough and weathered walls while floodlight brings out its character and shows the history of the building. LED-lighting can be changed according to the nights' atmosphere. In this way the small stage appears to be floating in space. Together with the enlarged void over the entrance this creates a spatial character while looking up as you arrive.

The underside of the small hall appears as the hull of a ship while it is clad in thin metal perforated sheets that also serve an acoustic purpose. All service points, clad in brushed oak

The enlarged void over the entrance creates a spatial character while looking up as you arrive.　扩张的入口在进门处凸显了空间特色。

1. Entrance
2. Exit
3. Ticket counter
4. Bar
5. Non smoking lounge
6. Smoking lounge
7. Cloakroom
8. Lockers
9. Vending / cash machine
10. Multiple mobile counters
11. Toilets
12. Stair towards small and big stage
13. Stair towards small stage
14. Stair towards offices

1. 入口
2. 出口
3. 售票处
4. 吧台
5. 禁烟休息室
6. 吸烟休息室
7. 衣帽间
8. 储物柜
9. 自动售货机 / 取款机
10. 多功能移动柜台
11. 卫生间
12. 通往大小舞台的楼梯
13. 通往小舞台的楼梯
14. 通往办公室的楼梯

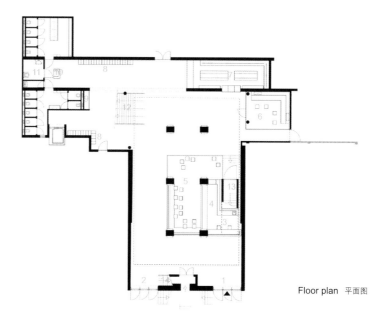

Floor plan　平面图

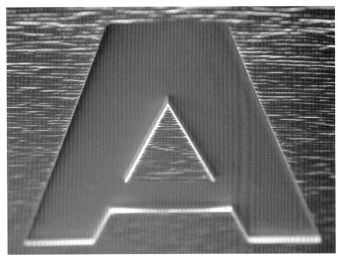

Brushed oak panels have recessed lettering showcasing their function while floodlight brings out its character and enhances readability.

橡木上凹陷的文字显示了各个服务点的功能,泛光灯突出了文字的特色和可读性。

panels have recessed lettering showcasing their function while floodlight brings out its character and enhances readability. The glass wall separating the bar from the foyer contains all the names of the sponsors. The glass walls of the smoking room are adorned with lyrics from the band DI-RECT.

In contrast with the rough walls and steel ceiling, comfortable oak-wood benches have been designed as seating arrangement. Air climatisation machinery has all been integrated within these benches hiding them from sight. Antique theatre spots provide functional lighting needed in specific areas. The new layout, solid use of materials and spatial lighting has transformed the foyer into a comfortable lounge. "'Het Paard van Troje' has grown up!" commented a regular visitor of the concert hall. The Hague can be proud of its new public living room.

拥有40年历史的海牙特洛伊木马音乐厅在2012年夏季进行了全新的翻修。两家来自海牙的工作室——芭芭拉·沃思工作室与椴树工作室受雇将音乐厅的前厅改造为城市公共大厅。设计以"原生态闪耀设计"为主题,紧贴音乐厅的公共职能。夜间活动不仅是参加盛大的活动,也需要私人空间。

自从10年前改造之后,特洛伊木马音乐厅已经成为闻名荷兰和全世界的著名音乐厅。音乐厅的走廊、楼梯和大厅由OMA事务所进行了重新设计。在这10年中,观众的需求有所变化,音乐厅希望创造一个舒适的酒吧和休息区,供人们在音乐厅外小酌或享用点心。为了庆祝成立40周年,特洛伊木马音乐厅在2012年夏季进行了全面改造。

当前,文化项目的融资也采用了新形势。特洛伊木马音乐厅在2012年春季进行了特殊形式的公开发售。翻修工程的部分资金来源于群众集资。通过入股,公众不仅可以参与到翻修工程中,还能将自己的名字刻在玻璃上,成为音乐厅的一部分。海牙著名乐队DI-RECT就以这种形式获得了吸烟室的命名权。

设计最大的特色就是将工业材料与简洁的外观结合起来。原始设计的强势元素被巧妙地加以强调,凸显了它的粗放空间特色。混凝土结构被涂上了浅色涂料,非承重墙由清水桦木包裹,而地面则以棕色复合材料铺成。

酒吧和吸烟室的私密感通过烟熏橡木地板与门厅其他空间的豪放氛围形成对比。浅黄色拱面支撑着小舞台。电线全部裸露在粗糙的风化墙壁上,泛光灯凸显了建筑的历史。LED照明可以根据夜晚的氛围进行变化,让舞台宛如悬浮在空中。这与扩张的入口共同营造了空间特色。

小型音乐厅的下方看起来像轮船的外壳一样,由冲孔薄板包裹,起到了特殊的音效作用。各个服务点被橡木板包围,上面书写着功能文字,并且以泛光灯突出了文字的特色和可读性。玻璃墙将酒吧与大厅隔开,上面刻有所有项目赞助人的名字。吸烟室的玻璃墙上装饰着DI-RECT乐队的歌词。

与粗糙的墙壁和钢铁天花板形成鲜明对比,舒适的橡木长椅布置整齐。这些长椅内结合了隐藏的空调系统。古老的剧院灯光为特殊场合提供了功能照明。全新的布局、材料与空间照明的合理运用将音乐厅的门厅改造成了舒适的休息大厅。一位音乐厅的常客评论道:"'特洛伊木马'成长了!"全新的公共大厅是海牙的骄傲。

The old cloakroom made place for the new pub. 旧衣帽间被改造成了新酒吧。

The benches are made of small oak laths and the floor of dark brown composite. 长椅由橡木条制成，而地板采用了深棕色复合材料。

The underside of the small hall appears as the hull of a ship while it is clad in thin metal perforated sheets that also serve an acoustic purpose.

The intimate living room character of the bar and smoking room, contrasts with the robust atmosphere of the rest of the foyer through the use of smoked oak floorboards.

小厅的下方看起来像轮船的外壳一样,由冲孔薄板包裹,起到了特殊的音效作用。

具有私密特征的酒吧和吸烟室通过烟熏橡木地板与门厅其他空间的豪放氛围形成对比。

In the smoking room and the pub air climatisation machinery has been integrated within these benches hiding them from sight.

Cables have all been directed over the rough and weathered walls while floodlight brings out its character and shows the history of the building. LED-lighting can be changed according to the nights' atmosphere.

吸烟室和酒吧中的空调机被隐藏在长椅之中。

电线全部裸露在粗糙的风化墙壁上，泛光灯凸显了建筑的历史。LED 照明可以根据夜晚的氛围进行变化，让舞台宛如悬浮在空中。

The glass wall separating the bar from the foyer contains all the names of the sponsors.

玻璃墙将酒吧与大厅隔开,上面刻有所有项目赞助人的名字。

The glass walls of the smoking room are adorned with lyrics from the band DI-RECT.

吸烟室的玻璃墙上装饰着DI-RECT乐队的歌词。

Antique stadion lights and theatre spots provide functional lighting needed in specific areas

古老的剧院灯光为特殊场合提供了功能照明

Index 索引

ACXT Architects
Tel: +34 91 444 11 63
Fax: +34 91 447 31 87

Andre Kikoski Architect
180 Varick Street, Suite 1316 New York, New York 10014
Tel: 212 627 0240
Fax: 212 627 0242

Brinkworth
6 Ellsworth Street
London E2 0AX
United Kingdom
Tel: +44(0)20 7613 5341
Fax: +44(0)20 7739 8425

colab studio, llc
1614 East Cedar Street, Tempe . Arizona 85281
Tel: 480 326 0541 voice
Fax: 480 967 2440 fax

DiPPOLD Innenarchitektur GmbH
Oskar-von-Miller-Ring 31
80333 München
Tel: +49 (0) 89 - 28 70 21 46
Fax: +49 (0) 89 - 28 70 21 47

Doyle Collection Co.,Ltd
1-20-3-302, Ebisu-minami, Shibuya-ku, Tokyo, Japan
Zip: 150-0022
Tel: +81-3-5734-1508
Fax: +81-3-5734-1509

GRITTI ARCHITETTI
via S. Antonino 11
24122 Bergamo (BG), Italy
Tel: +39 035 21 31 14

Maarten Baas
Baas & Den Herder BV
Rosmalensedijk 3
5236 BD 's Hertogenbosch (Gewande)
The Netherlands

MInas Kosmidis
Filellinon 55
Panorama, 552 36
Thessaloniki, Greece
Tel: +30 2310 344651
fax: +30 2310 344656

M+R interior architecture
Aalsterweg 230
5644 RK Eindhoven

The Netherlands
P.O. Box 1115
5602 BC Eindhoven
Tel: +31 40 21 37408
Fax: +31 40 21 37409

Project Orange
1st Floor
Cosmopolitan House
10A Christina Street
London EC2A 4PA
United Kingdom
Tel: +44 (0)20 7739 3035
Fax: +44 (0)20 7739 0103

Rockwell Group
5 Union Square West, 8th Floor
New York, NY, 10003
Tel: 212 463 0334
Fax: 212 463 0335

SHH
1 Vencourt Place
Ravenscourt Park
Hammersmith
LondonW6 9NU
United Kingdom

Tel: +44 (0)20 8600 4171
Fax: +44 (0)20 8600 4181

SMS Architects
17848 Sky Park Circle, Suite B
Irivine, California 92614
Tel: 949-757-3240

Softroom
31–37 Hoxton Street London N1 6NL
Tel: + 44 (0)20 7408 0864

Studio Barbara Vos
Ateliercomplex 'De Besturing'
Saturnusstraat 89
2516 AG Den Haag, NL
Tel: +31 (0)6 2478 53 66

Studio Odile Decq
11 rue des Arquebusiers
75003 Paris, France
Tel: 33 1 42 71 27 41
Fax: 33 1 42 71 27 42

Very Good and Proper
info@verygoodandproper.co.uk

图书在版编目（CIP）数据

轻食餐厅 /（英）海斯编；鄢格译. -- 沈阳：辽宁科学技术出版社，2014.3
ISBN 978-7-5381-8441-9

Ⅰ. ①轻… Ⅱ. ①海… ②鄢… Ⅲ. ①餐馆-室内装饰设计-图集 Ⅳ. ①TU247.3-64

中国版本图书馆CIP数据核字(2013)第319213号

出版发行：辽宁科学技术出版社
　　　　　（地址：沈阳市和平区十一纬路29号 邮编：110003）
印　刷　者：利丰雅高印刷（深圳）有限公司
经　销　者：各地新华书店
幅面尺寸：215mm×285mm
印　　张：14
插　　页：4
字　　数：50千字
印　　数：1～1200
出版时间：2014年 3 月第 1 版
印刷时间：2014年 3 月第 1 次印刷
责任编辑：陈慈良　于峰飞
封面设计：何　萍
版式设计：何　萍
责任校对：周　文
书　　号：ISBN 978-7-5381-8441-9
定　　价：228.00元
联系电话：024-23284360
邮购热线：024-23284502
E-mail: lnkjc@126.com
http://www.lnkj.com.cn
本书网址：www.lnkj.cn/uri.sh/8441